Pelican Books

The Flamingo's Smile

Stephen Jay Gould grew up in New York City. He graduated from Antioch College and received his Ph.D. from Columbia University in 1976. Since then he has been on the faculty of Harvard University. He considers himself primarily a palaeontologist and an evolutionary biologist, though he teaches geology and the history of science as well. A frequent and popular speaker on the sciences, his published work includes *Ontogeny and Phylogeny*, a scholarly study of the theory of recapitulation; *The Mismeasure of Man* (Pelican, 1983), winner of the National Book Critics' Circle Award for 1982; the popular collections of essays *Ever Since Darwin: Reflections in Natural History* (Pelican, 1980) which received great acclaim: 'Unreservedly, they are brilliant' – *New Scientist*; *The Panda's Thumb: More Reflections in Natural History* (Pelican, 1983), which won the 1981 American Book Award for Science; *Hen's Teeth and Horse's Toes: Further Reflections in Natural History* (Pelican, 1984); *Time's Arrow, Time's Cycle* (Pelican, 1988); and *An Urchin in the Storm* (1988).

STEPHEN JAY GOULD

THE
FLAMINGO'S
SMILE

Penguin Books

For Deb for everything

PENGUIN BOOKS

Published by the Penguin Group
27 Wrights Lane, London w8 5tz, England
Viking Penguin Inc., 40 West 23rd Street, New York, New York 10010, USA
Penguin Books Australia Ltd, Ringwood, Victoria, Australia
Penguin Books Canada Ltd, 2801 John Street, Markham, Ontario, Canada l3r 1b4
Penguin Books (NZ) Ltd, 182–190 Wairau Road, Auckland 10, New Zealand

Penguin Books Ltd, Registered Offices: Harmondsworth, Middlesex, England

First published in the USA by W. W. Norton 1985
First published in Great Britain in Pelican Books 1987
Reprinted 1988

Made and printed in Great Britain by
Richard Clay Ltd, Bungay, Suffolk

Contents

8 | EXTINCTION AND CONTINUITY

The
Flamingo's
Smile

Prologue

IN THE MEDIEVAL GLASS of Canterbury Cathedral, an angel appears to the sleeping wise men and warns them to go straight home, and not return to Herod. Below, the corresponding event from the Old Testament teaches the faithful that each moment of Jesus' life replays a piece of the past and that God has put meaning into time—Lot turns round and his wife becomes a pillar of salt (the white glass forming a striking contrast with the glittering colors that surround her). The common theme of both incidents: don't look back.

The Flamingo's Smile is my fourth volume of essays from monthly columns in *Natural History Magazine;* it also contains my hundredth contribution to a genre that I once considered both more ephemeral and impossible to sustain. Thus, I will also break Lot's injunction, hope for a sweeter fate, and look back upon the previous volumes.

One brand of Scotch often graces *New Yorker* back covers with its claim that Angus Mac-somebody-or-other (and ancestors of that ilk) have been throwing the caber on the same field since 1367, give or take a few years. "Some things never change," the bottom line (literally) proclaims. Some things better change (however difficult under punctuated equilibrium), if only to allay boredom, but fundamental themes (like a successful blend) should revel in persistence. If my volumes work at all, they owe their reputation to coherence supplied by the common theme of evolutionary theory. I have a wonderful advantage among essayists because no other theme so beautifully encompasses both the particulars that fascinate and the generalities that instruct.

Evolution is one of the half-dozen shattering ideas that science has developed to overturn past hopes and assumptions, and to enlighten our current thoughts. Evolution is also more personal than the quantum, or the relative motion of earth and sun; it speaks directly to the questions of genealogy that so fascinate us—how and when did we arise, what are our biological relationships with other creatures? And evolution has built all those creatures in stunning variety—an endless source of delight (though not the reason for their existence!), not to mention essays.

To map the changes within this persistence, I reread the prefaces to my other volumes and found a coordinating theme, linked to times of composition, for each. *Ever Since Darwin,* as a first attempt, presented the basics of evolutionary theory as a comprehensive world view with implications for a political world (of years just following the Vietnam War) that treated human diversity more generously. *The Panda's Thumb* highlighted a series of debates (about rates and results) that arose among professional evolutionists and imparted renewed vigor and range to "this view of life." *Hen's Teeth and Horse's Toes* appeared in the shadow of resurgent Yahooism—so-called "creation science" as preached by Falwell and company—and required a gentle defense of both the veracity and humanity of evolution.

The Flamingo's Smile has a different kind of trigger—a specific discovery with cascading implications. It now seems, to use the favored jargon of the profession, "highly probable" that an errant asteroid or shower of comets provoked the great Cretaceous extinction (dinosaur death knell and, conversely, the Introit for our own evolution). Moreover, such quintessentially fortuitous and episodic restructurings of life have occurred several times, perhaps even on a regular cycle of some 25–30 million years. The particulars are striking (pun intended, I suppose), but the general implications are even more arresting, and beautifully coincident with persistent themes that infest all my columns—the meaning of pattern in life's history (partly random and, in any case, not designed for us or towards us); the social implications of scientific assaults upon pervasive biases of Western thought (my favorite four horsemen of progress, determinism, gradualism, and adaptationism—all severely questioned by the impact theory of mass extinction). At the center stands the one theme that transcends even evolution itself in generality—the nature of history. *The Flamingo's Smile* is about history and what it means to say that life is the product of a contingent past, not the inevitable and predictable result of simple, timeless laws of nature. Quirkiness and meaning are my two not-so-contradictory themes.

All this sounds awfully tendentious and may lead readers to fear that potential pleasure has been sacrificed on a bloated altar of imagined importance (my volumes have become progressively longer for an unchanging number of essays—a trend more regular than my mapped decline of batting averages from essay 14, and a warning signal of impending trouble if continued past a limit reached, I think, by this collection). My potential salvation in the face of admitted egotism must remain an unswerving commitment to treat generality only as it emerges from little things that arrest us and open our eyes with "aha"—while direct, abstract, learned assaults upon generalities usually glaze them over. Even my most grandiloquent essay (emphatically not my best)—number 29 on continuity itself—arose as a gloss on a small observation: the mingling of sacred and profane in the iconography of Pio Quatro's Palace in the Vatican.

I placed my essays on reversals and boundaries (part 1) at the beginning because they best exemplify this style of letting generality cascade out of particulars—three essays on inversions of general expectations (flamingos that feed upside down; female insects that supposedly eat their mates after copulation; flowers and snails that change from male to female, and sometimes back again); and two on continua and the problem of boundaries in nature (are Portuguese men-of-war individuals or colonies, are Siamese twins one person or two). Each essay is both a single long argument and a welding together of particulars.

Throughout most of Europe, the communication of science to a general audience has been viewed as part of humanism, as an honorable intellectual tradition stretching from Galileo, who wrote in Italian to bring science beyond the Latin confines of church and university, to Thomas Henry Huxley, who was as fine a literary stylist as many a great Victorian novelist, to J.B.S. Haldane and Peter Medawar in our own times. In America, this worthy activity has been badly confused with the worst aspects of journalism, and "popularization" has become synonymous in some quarters with bad, simplistic, trivial, cheapened, and adulterated. I follow one cardinal rule in writing these essays—no compromises. I will make language accessible by defining or eliminating jargon; I will not simplify concepts.

I can state all sorts of highfalutin, moral justifications for this approach (and I do believe in them), but the basic reason is simple and personal. I write these essays primarily to aid my own quest to learn and understand as much as possible about nature in the short time allotted. If I play the textbook or TV game of distilling the already known, or shearing away subtlety for bare bones accessible in the vulgar sense (no return work required from consumers), then what's in it for me?

All these essays are based on original sources in their original languages: none are direct reports from texts and other popular summaries. (The propagation of error, by endless transfer from textbook to textbook, is a troubling and amusing story in its own right—a source of inherited defect almost more stubborn than inborn errors of genetics.) My errors are my errors.

These essays, in this light, fall into three categories. Most are exercises in personal scholarship. Some reach new interpretations (at least to me): I think that my reading of Tyson as a conservative supporter of the chain of being and not as an innovative pioneer of evolution resolves the disparities between his text and the usual analyses (essay 17); I found that Wells's first statement of natural selection is not so consonant with Darwin's later version as most commentators have held (essay 22); although Kinsey's previous life as a wasp taxonomist has not been hidden, I don't think that its intimate intellectual connection with his sex research had been properly traced (essay 10—I suspect that such a treatment required a professional taxonomist working from the wasp's end, and not a psychologist proceeding backwards). Other essays represent discoveries of sorts based on new data. It may not be clear from the jocular tone of the essay, but more work (of the dumb tabulatory type —a kind of perverse or benumbing pleasure in itself) lies hidden in my chart of low batting averages through time (essay 14) than in many fancy analyses flaunted in my technical papers on land snails. (Each of my volumes contains an essay in lighter relief—allometry of cathedrals in *Ever Since Darwin*, neoteny of Mickey Mouse in *The Panda's Thumb*, shrinking Hershey bars in *Hen's Teeth and Horse's Toes*, extinction of .400 hitting herein. I often insist that these are serious pieces, and I really mean it—though I will be absolutely crushed if you don't chuckle. I almost regret the chosen illustration, the history of batting averages, in essay 14, because its general theme—a plea for considering systems rather than abstracted parts—should be part of the fun, not lost in it.)

In a second category, I report the discoveries or interpretations of friends and colleagues, but I embed them in a personal theme. I use Iltis's theory on the origin of corn (essay 24) to illustrate the most difficult and important evolutionary concept of homology; the discovery of the conodont animal (essay 16) becomes a handle for discussing what may be the basic (but underappreciated) pattern of life's history—reduction in diversity of morphological designs, with marked expansion among survivors.

The third category sets general themes that need an airing, but searches for quirky or unusual particulars for their illustration. Essays 4 and 5 are an experiment—the same theme twice, with radically different illustrations. I discuss reductionism via the tragic life of E.E. Just (essay 25) and the numerology of antiquated pre-Darwinian taxonomies (essay 13). I temper *the* nature of science with some funny ideas about dinosaurs (essay 28) and a plea for Mr. Gosse (essay 6), who argued that as God created animals with feces in their intestines, so too did he make the earth with coprolites (fossil excrement) in its strata.

I also hope that the arrangement of essays into categories aids my larger purpose by emphasizing, via juxtaposition, the themes expressed in essays taken separately. In making three statements about the chain of being (17–19), I try to show how the unavoidable embedding of science in culture acts both as a constraint (in defending unwarranted prejudice as certified knowledge, with tragic consequences for individual lives—essay 19 on the Hottentot Venus), and as a fruitful prod to new discovery that may influence culture in return (the chain of being guided Tyson to some remarkable data about the anatomy of chimpanzees—essay 17).

My profession embodies one theme even more inclusive than evolution—the nature and meaning of history. History employs evolution to structure biological events in time. History subverts the stereotype of science as a precise, heartless enterprise that strips the uniqueness from any complexity and reduces everything to timeless, repeatable, controlled experiments in a laboratory. Historical sciences are different, not lesser. Their methods are comparative, not always experimental; they explain, but do not usually try to predict; they recognize the irreducible quirkiness that history entails, and acknowledge the limited power of present circumstances to impose or elicit optimal solutions; the queen among their disciplines is taxonomy, the Cinderella of the sciences. As I wrote *Hen's Teeth and Horse's Toes,* I watched with almost detached amusement as history slowly emerged in the forefront of my concerns. It has spread through this volume like a transposon. The flamingo's smile (like the panda's thumb) is its synecdoche—a quirky structure, constrained by a different past, and cobbled together from parts available.

Essay 12, on contingent vs. necessary facts, may be my most direct statement about history, but this subject laces the entire volume. I pondered for a long time about my hundredth essay, for I thought that it should epitomize my efforts. I wrote on the importance of taxonomy, as applied to the West Indian land snails that serve as a focus for my technical research in biology. Taxonomy, the most underappreciated of all sciences, is the keystone of historical disciplines. Part 3 celebrates taxonomy in many guises. Other essays also discuss the methods of history—essay 24 on homology as a guide to descent; essays 4 and 5 on the meaning of boundaries in a world of continua.

Several sections treat the patterns that history produces by its mandated process of evolution—part 4 on trends in the history of life (and some smaller systems); part 8 on extinction as so much more than a negative force; part 7 on life here on earth, and the predictions that history permits about life elsewhere (the limits of contingency again, I fear, rather than the blueprints for ET). Finally, if history matters and science cannot be reduced to robotic experiment, then the interplay of science with culture and personality is no mere nuisance, but a spur to creativity and a key to understanding. Part 5 treats the interplay for themes of human evolution. Part 2 preaches respect for fine scientists who were misunderstood or ridiculed by the arrogant approach that regarded history only as a repository of error and, thereby, a source of moral instruction. I confess a particular fondness for essay 5 and its poignant subject.

If Alvarez's asteroid was the external prod to cohesion, this book has an internal theme as well. It is no particular secret that I have spent the last few years fighting cancer. My disease was diagnosed just a week after the last volume went to the printers. This book becomes, therefore, a kind of *roman à clef* (complete, I hope) to a personal odyssey. Essay 19, "The Hottentot Venus," was the first piece I wrote as a member of this largest involuntary club—and its last line was my *touché*. When arranged in their order of appearance in *Natural History*, these essays may trace an emotional journey (though I do not choose to pursue the analysis). I will only say that some essays are spare in their exegetical style of commentary on single historical texts (for I couldn't reach the libraries for my usual ramblings, and several nights with a beautiful old book were such a solace), while others are downright baroque in their patchwork of detail (my simple joy in being able once again).

I dare not even try to express my thanks to those who supported me through all this; the words do not exist in any language. But to those who know me only through these essays, and who took time to tell me that they cared, my special gratitude; it mattered in the most palpable way. I dwelled on many things—that I simply had to see my children grow up, that it would be perverse to come this close to the millennium and then blow it. I hope that it won't sound corny if I thank nature too—in the context of the plodding regularity of these essays. Who can surpass me in the good fortune they supply; every month is a new adventure—in learning and expression. I could only say with the most fierce resolution: "Not yet Lord, not yet." I could not dent the richness in a hundred lifetimes, but I simply must have a look at a few more of those pretty pebbles.

1 | Zoonomia (and Exceptions)

1 | The Flamingo's Smile

BUFFALO BILL played his designated role in reducing the American bison from an estimated population of 60 million to near extinction. In 1867, under a contract to provide food for railroad crews, he and his men killed 4,280 animals in just eight months. His slaughter may have been indiscriminate, but the resulting beef was not wasted. Other despoilers of our natural heritage killed bison with even greater abandon, removed the tongue only (considered a great delicacy in some quarters), and left the rest of the carcass to rot.

Tongues have figured before in the sad annals of human rapacity. The first examples date from those infamous episodes of gastronomical gluttony—the orgies of Roman emperors. Mr. Stanley, Gilbert's "modern major general," could "quote in elegiacs all the crimes of Heliogabalus" (before demonstrating his mathematical skills, in order to cadge a rhyme, by mastering "peculiarities parabolous" in the study of conic sections). Among his other crimes, the licentious teen-aged emperor presided at banquets featuring plates heaped with flamingo tongues. Suetonius tells us that the emperor Vitellius served a gigantic concoction called the Shield of Minerva and made of parrot-fish livers, peacock and pheasant brains, lamprey guts and flamingo tongues, all "fetched in large ships of war, as far as from the Carpathian sea and the Spanish straights."

Lampreys and parrot fishes (though not without beauty)

23

have rarely evoked great sympathy. But flamingos, those elegant birds of brilliant red (as their name proclaims), have inspired passionate support from the poets of ancient Rome to the efforts of modern conservationists. In one of his most poignant couplets, Martial castigated the gluttony of his emperors (circa 80 A.D.) by speculating about different scenarios, had the flamingo's tongue been gifted with song like the nightingale's, rather than simple good taste:

Dat mihi penna rubens nomen; sed lingua gulosis
Nostra sapit: quid, si garrula lingua foret?

(My red wing gives me my name, but epicures regard my tongue as tasty. But what if my tongue could sing?)

Most birds have skinny pointed tongues, scarcely fit for an emperor, even in large quantities. The flamingo, much to its later and unanticipated sorrow, evolved a large, soft, fleshy tongue. Why?

Flamingos have developed a surpassingly rare mode of feeding, unique among birds and evolved by very few other vertebrates. Their bills are lined with numerous, complex rows of horny lamellae—filters that work like the whalebone plates of giant baleen whales. Flamingos are commonly misportrayed as denizens of lush tropical islands—something amusing to watch while you sip your rum and coke on the casino veranda. In fact, they dwell in one of the world's harshest habitats—shallow hypersaline lakes. Few creatures can tolerate the unusual environments of these saline deserts. Those that thrive can, in the absence of competitors, build their populations to enormous numbers. Hypersaline lakes therefore provide predators with ideal conditions for evolving a strategy of filter feeding—few types of potential prey, available in large numbers and at essentially uniform size. *Phoenicopterus ruber,* the greater flamingo (and most familiar species of our zoos and conservation areas in the Bahamas and Bonaire), filters prey in the predominant range of an inch or so—small mollusks, crustacea, and insect larvae, for example. But *Phoeniconaias minor,* the lesser flamingo, has filters so dense and efficient that they segregate cells of blue-green algae and diatoms with diameters of 0.02 to 0.1 mm.

Flamingos pass water through their bill filters in two ways (as documented by Penelope M. Jenkin in her classic article of 1957): either by swinging their heads back and forth, permitting the water to flow passively through, or by the usual and more efficient system that inspired the Roman gluttons—an active pump maintained by a large and powerful tongue. The tongue fills a large channel in the lower beak. It moves rapidly back and forth, up to four times a second, drawing water through the filters on the backwards pull and expelling it on the forward drive. The tongue's surface also sports numerous denticles that scrape the collected food from the filters (just as whales collect krill from their baleen plates).

The extensive literature on feeding in flamingos has highlighted the unique filters—and often neglected another, intimately related, feature equally remarkable and long appreciated by the great naturalists. Flamingos feed with their heads upside down. They stand in shallow water and swing their heads down to the level of their feet, subtly adjusting the head's position by lengthening or shortening the s-curve of the neck. This motion naturally turns the head upside down, and the bills therefore reverse their conventional roles in feeding. The anatomical upper bill of the flamingo lies beneath and serves, functionally, as a lower jaw. The anatomical lower bill stands uppermost, in the position assumed by upper bills in nearly all other birds.

With this curious reversal, we finally reach the theme of this essay: Has this unusual behavior led to any changes of form and, if so, what and how? Darwin's theory, as a statement about adaptation to immediate environments (not general progress or global direction), predicts that form should follow function to establish good fit for peculiar life styles. In short, we might suspect that the flamingo's upper bill, working functionally as a lower jaw, would evolve to approximate, or even mimic, the usual form of a bird's lower jaw (and vice versa for the anatomical lower, and functionally upper, beak). Has such a change occurred?

The enigmatic smile of a swan—or is it?

Nature harbors a large suite of oddities so special that we scarcely know what to predict. But, in this case, we encounter a precise reversal of anatomy and usual function—leading to a definite expectation: upside-down animals should reorient the form of their bodies to a new function when current behavior and conventional anatomy conflict.

We may begin by sparing the usual pontification (but only for a while) and looking at a picture. If this picture excites a vague feeling of familiarity slightly awry, your perceptions are acute, but ride with me for a while.

We seem to see a long-necked swan with a broad smile. But look carefully, for details betray this impossible beast. Its mouth opens *above* the eyes; its feathers run the wrong way; and where are its legs? I now show you the celebrated original in its proper orientation (and with the legs restored)—the flamingo from J.J. Audubon's *Birds of America,* and a sure entry on anyone's hit parade of most famous pictures in natural history.

This dramatic perceptual switch from happy swan to haughty flamingo recalls any standard item in the psychological arsenal of optical illusion—particularly the young well-dressed lady looking away who becomes the old hag in profile. In fact, any accurately executed picture of a flamingo produces the same jolting effect when viewed upside down (I have checked all historically important portraits)—and for an obvious reason. The jaws have evolved to fit their reversed function. The flamingo's upper jaw does look like a typical bird's lower bill, and we therefore see the upside-down flamingo not as an absurdity, but as an only slightly odd swan-like bird.

The morphological alterations extend far beyond the changes in external form that produce such a striking perceptual shift from upright flamingo to inverted "swan." But note first the peculiar curve of the beak itself. The flamingo's bill projects out from its face, but then makes a sharp angular turn, producing the pronounced hump that looks like a trough (and works like one) when inverted for feeding. Some Near Eastern peoples call flamingos "camels of the sea," not because the inclined bill recalls the hump on a camel's back, but because it mimics the bend of the nose that imparts an inappropriate (but unshakable) impression of haughtiness to both animals (see my essay on the history of Mickey Mouse and the messages accidentally conveyed by facial features of animals—essay 9 in *The Panda's Thumb*). Turned upside down, the hump becomes a grin as a smiling "swan" replaces the haughty flamingo.

The bills are elaborately adapted to their reversed roles, not simply bent in the middle for proper reorientation. First, relative sizes have been rearranged to complement

The famous flamingo, FROM J.J. AUDUBON'S *Birds of America*.

the shapes. The upper bill is small and shallow, the lower deep and massive. (In most birds, the smaller lower bill moves up and down against the larger upper beak.) Second, the flamingo's lower bill (functionally upper in feeding) has evolved unusual rigidity. The bones of each half (or ramus, in technical parlance) are tightly fused, and the rami themselves are then bonded extensively to each other. The lower bill is massive and well braced. The tongue runs fore and aft in a deep trough cut into the lower jaw. (Remember that filter feeding serves as a coordinating theme for all these changes—the upside-down feeding posture, the attendant alteration in size and shape of the bills, and the fat tongue that once almost sealed the flamingo's fate.) Third, in most species of flamingo, the smaller upper jaw slots into a receiving space on the larger lower jaw, a reversal of the usual convention—lower jaw moving up and fitting into a larger upper bill.

These complex, coordinated changes make a persuasive case, but they leave a missing piece, recognized as the key to flamingo peculiarities ever since Menippus recorded the first preserved speculation nearly 300 years before Martial's plea: are movements also reversed to match the inversion of form?

In most birds (and mammals, including us), the upper jaw fuses to the skull; chewing, biting, and shouting move the mobile lower jaw against this stable brace. If reversed feeding has converted the flamingo's upper jaw into a working lower jaw in size and shape, then we must predict that, contrary to all anatomical custom, this upper beak moves up and down against a rigid lower jaw. The flamingo, in short, should feed by raising and lowering its upper jaw.

With great credit to the clear thinking of our finest naturalists, I noted with pleasure in my readings that this central question has been continually posed as paramount for more than 2,000 years—by scientists of many cultures and through all the vicissitudes of theory and practice that have marked the history of biology. Georges Buffon, the greatest of all synoptic naturalists, began his mid-eighteenth-century essay on flamingos by admitting the fame of their red color, while maintaining that the odd form of their beak posed a problem of even greater interest: "This fiery color is not the only striking character displayed by this bird. Its beak has an extraordinary form, the upper bill flattened and strongly bent at its midpoint, the lower thick and well set, like a large spoon." In short, and in his own lovely tongue, "une figure d'un beau bizarre et d'une forme distinguée." Then, tracing the question right back to Menippus, Buffon stated the *primum desideratum* of flamingo studies—"to know if, in this singular beak, it is (as many naturalists have said) the upper part that moves, while the lower remains fixed and motionless."

The first extensive and explicit commentary had been offered in 1681 by Nehemiah Grew, the great English naturalist (known primarily for his early microscopical studies of plants). Cataloguing the collections of the Royal Society— in his *Musaeum Regalis Societatis, or a catalogue and description of the natural and artificial rarities belonging to the Royal Society and preserved at Gresham Colledge, whereunto is subjoyned the comparative anatomy of stomachs and guts*—he encountered a lone flamingo (see figure) and stated: "that wherein he is most

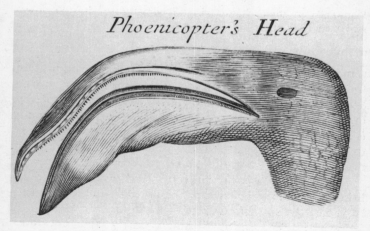

Phoenicopter's Head

Nehemiah Grew's flamingo, 1681. The illustration accompanying the first important proposal that flamingos feed by moving their upper jaw up and down against their lower. Look at this figure upside down as well. FROM N. GREW, MUSAEUM REGALIS SOCIETATIS, 1681. REPRINTED FROM NATURAL HISTORY.

Flamingos in their characteristic feeding pose—upside down.
PHOTO BY D. PURCELL.

remarkable, is his bill." Grew suspected that the oddities of
the bill would all be resolved if the upper beak moved
against a stationary lower jaw. He stated that the "shape and
bigness of the upper beak (which here, contrary to what it
is in all other birds that I have seen, is thinner and far less
than the nether) speaks it to be more fit for motion and to
make the appulse and the nether to receive it."

The question was not fully resolved until Jenkin published her comprehensive paper in 1957—affirming with
hard data the suspicions and good judgment of Menippus,
Grew, and Buffon. In fact, flamingos (along with many other
birds) have developed a highly mobile ball and socket joint
between upper and lower jaws. The beaks therefore have
great mobility, and each can move independently. In preen-
ing, either the upper or lower jaw may be opened and operated against the other. But, in feeding, the upper jaw usually

drops and raises against a stationary lower jaw—just as the great naturalists had always expected.

The flamingo's flip-flop is complete and comprehensive —in form *and* motion. The shapes are overturned by bending, the sizes reversed, the slotting inverted, the buttressing transposed. The action, too, is topsy-turvy. A peculiar reversal in behavior has engendered a complex inversion of form. Evolution as adaptation to particular modes of life— Darwin's vision—gains strength from an extreme test imposed by life upside down.

But do flamingos just provide a funny example, or do they symbolize a generality? What about other creatures that live upside down? Consider another animal of shallow West Indian waters, the inverted jellyfish, *Cassiopea xamachana* (the unorthodox trivial name honors the Native American designation for the island of Jamaica).

Cassiopea is an unconventional jellyfish in many ways. It grows neither marginal tentacles nor central mouth. Instead, eight fleshy and complexly branched "oral arms" (so called because each contains a separate mouth) emerge from a short and stout central stalk, itself attached to a usual jellyfish umbrella with a difference (see figure—a reproduction of the classical lithograph from Mayer's 1910 monograph, *Medusae of the World*). The oral arms are crammed with symbiotic algal cells, a possible adaptive impetus for their elaborate branching (to provide light-capturing surfaces for the photosynthetic symbionts). Each oral arm harbors about forty oral vesicles—hollow sacs connected with the feeding canals and containing bags of nematocysts, or stinging cells, at their tips. The vesicles shoot their nematocysts at prey (mostly small crustaceans) in strings of mucus; the strings with their attached and paralyzed victims are then pulled into the oral mouths. (Yes, I was as amused as some of you by the redundant "oral mouth"—the zoological equivalent of pizza pie or AC current. This clumsy phrase arises as unfelicitous fallout from a prior decision to call the appendages "oral arms"—as a shortcut for "mouths of the oral arms.")

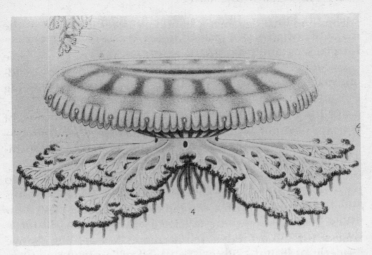

Cassiopea xamachana. Note concavity of bell's upper surface and the raised muscular ring. Figure reproduced as presented (in the ecologically wrong right-side-up position. FROM MAYER, 1910. REPRINTED FROM NATURAL HISTORY.

Cassiopea's unusual anatomy matches its unconventional orientation and style of life. Ordinary, self-respecting jellyfish swim actively with their umbrellas uppermost and their arms and tentacles below. *Cassiopea* lies stationary on the bottom of shallow ponds and coastal areas—upside down. The top of its umbrella hugs the sediment and the oral arms wave above, waiting for small crustacea to enter their orbit. Sailors at Fort Jefferson in the Tortugas, where *Cassiopea* lined the docks, called them "moss cakes." (Since *Cassiopea* can give a nasty sting, and since men in blue usually spice their language to match the stimulus, I wonder what the sailors really called them. But Mr. H.F. Perkins, writing in 1908 on the anatomy of *Cassiopea,* didn't choose to tell us.)

The umbrella of *Cassiopea* recalls the flamingo's jaw in its adaptation to reversed life. The umbrella's upper surface is smoothly convex in ordinary jellyfish, as hydrodynamic efficiency dictates. But the upper surface of *Cassiopea*'s umbrella (its functional lower surface for life upside down) is markedly *concave*—well suited to serve as a cupping device for gripping and holding the substrate.

Cassiopea has made a second intriguing change for its unusual reversed life. Most jellyfish move through the water by contracting rings of concentric muscles that circle the outer portion of the umbrella. In *Cassiopea,* one of these muscle rings has been raised and accentuated, forming a continuous circular band surrounding the inner concavity. This raised rim operates together with the concave surface to form an efficient suction cup that keeps the "head" of this jellyfish in its proper position on the bottom. (*Cassiopea* can still swim, albeit weakly and inefficiently, in the conventional manner. If dislodged from the bottom, it will turn over and swim for a few pulsations before settling down again on its head.) Some scientists have also suggested that the pulsating contractions of the concentric muscles, ordinarily used in swimming, serve other important functions in *Cassiopea*'s attached, upside-down position—maintaining connection with the substrate by pushing the animal down and moving water currents with potential prey towards the oral arms. But these reasonable proposals have not been properly tested.

Thus, flamingos and *Cassiopea*—two animals that could scarcely differ more in design and evolutionary history—share the common feature of feeding upside down. As a general message amidst the particulars, they have both redesigned conventional anatomy to match reversed life style. The flamingo's upper bill has changed radically—in size, shape, and motion—to look and work like the lower beak of most birds. The structural top of *Cassiopea*'s umbrella has inverted its shape, all the better to work properly as an ecological bottom.

Adaptation has a wonderful power to alter an anatomical design, widespread and stable among thousands of species, for the reversed requirements of an odd life style assumed by one or a few aberrant forms. Yet, we should not conclude that Darwinian adaptation to local environments has unconstrained power to design theoretically optimum shapes for all situations. Natural selection, as a historical process, can only work with material available—in these cases, the conventional designs evolved for ordinary life. The resulting imperfections and odd solutions, cobbled together from parts on hand, record a process that unfolds in time from unsuited antecedents, not the work of a perfect architect creating *ab nihilo*. *Cassiopea* co-opts a band of muscles ordinarily used in swimming and forms a raised rim to grasp the substrate. Flamingos bend their bill in a curious hump as the only topological solution to a new orientation.

These adaptations to life upside down are not just funny facts. They help us to comprehend the solution to a major, and classical, dilemma in evolutionary theory (hence my decision to unite them in this essay). We can easily understand how flamingos and *Cassiopea* work; their unusual features do fit them for their unconventional lives. But how do these odd structures arise if evolution must proceed through intermediate steps (no one will seriously suggest that the first proto-flamingo turned its head upside down and then produced offspring with a complete set of complex adaptations to reversed life).

In pre-Darwinian years of the early nineteenth century, when evolution was new, and when early exponents of such a radical idea were trying to work out its ramifying implications, two schools emerged and carried out an interesting (and largely forgotten) struggle until Darwin resolved their debate. Both sides admitted the good fit that usually exists between form and function—adaptation in its static, non-historical meaning. Structuralists, like Etienne Geoffroy Saint-Hilaire argued that form must change first and then find a function. Functionalists, like Jean Baptiste Lamarck, held that organisms must first adopt a different mode of life to trigger some sort of pressure for a subsequently altered form.

The nature of this "pressure" inspired another famous (and better remembered, but no more important) debate. Lamarck held that organisms respond creatively to the needs imposed by their environments and then pass the resulting changes directly to offspring—"inheritance of acquired characters" in the usual jargon. Darwin argued that environments do not impose their adaptive requirements directly. Rather, those organisms that vary, by good fortune, in directions better suited to local environments leave more surviving offspring by a process of natural selection.

Since Darwin won this argument about the nature of signals that pass from environment to organism, Lamarck has been eclipsed and still, despite many efforts by historians to set the record straight, suffers from an imposed reputation as a loser not to be taken seriously for any of his ideas.

But Lamarck had the right answer (the same as Darwin's) to the larger dispute between structuralists and functionalists. (He only proposed the wrong mechanism for how environment gets its message to organisms.) Geoffroy's structuralist solution poses an obvious dilemma. If structure changes first, according to unknown "laws of form," and then finds the environment best suited to its altered state, how can precise adaptation arise? We might allow that some very basic and general changes could precede any functional meaning or advantage—an animal might, for example, get larger and then exploit the inherent advantages of increased size. But can we seriously believe that something so complex, so multifarious, and so intimately suited for an unusual ecology as the flamingo's bill might arise before the fact and without relationship to its usefulness—permitting the flamingo to discover only later how nicely such a beak worked upside down?

Lamarck's functionalist solution has an elegant simplicity accepted by nearly all evolutionists today (but usually attributed to Darwin, who also supported it. However much I revere Darwin, I want to advance a plea for recognizing this basic principle as Lamarck's primary contribution. It does not appear as an incidental footnote in Lamarck's *Philosophie zoologique* of 1809, but as a central theme of his book. Lamarck knew exactly what he was arguing and why.). Lamarck simply recognized that change of behavior must precede alteration of form. An organism enters a new environment with its old form suited to other styles of life. The behavioral innovation establishes a discordance between new function and inherited form—an impetus to change (by creative response and direct inheritance for Lamarck, by natural selection for Darwin). The protoflamingo first inverts its normal bill—and it doesn't work very well. The proto-*Cassiopea* turns over, but its convex umbrella doesn't clutch the substrate. Lamarck wrote:

> It is not the shape either of the body or its parts, which gives rise to the habits of animals and their mode of life; but it is, on the contrary, the habits, mode of life, and all the other influences of the environment, which have in course of time built up the shape of the body and of the parts of animals.

The direct evidence for Lamarck's solution cannot emerge from such "completed" adaptations as the flamingo's beak or *Cassiopea*'s umbrella—though the inference even here becomes quite compelling (for why should flamingos, uniquely among birds, develop such a peculiar beak if not to exploit their chosen, odd environment). We must catch the process at its beginning stages—by finding upside down animals that have already altered their behavior, but not their form.

African catfishes of the family Mochokidae include several species that characteristically swim upside down (see G. Sterba, in bibliography). Behavior has already changed radically, and we even have good hints about the triggers in some cases. (*Synodontis nigriventris,* for example, eats algae by grazing the undersides of leaves on water-dwelling plants.) But form has altered scarcely, if at all. A few species have reversed the usual pattern of cryptic coloration for fish swimming near the surface. The light bellies of most fish render them invisible to predators looking up through the water into sunlight above. But *S. nigriventris,* as its name (black belly) implies, is dark on its anatomical underside, and light on its structural top. Since this fish swims upside down, the light side lies below, as usual. Yet, beyond this switch in color, most upside-down mochokids look just like their upright relatives. Size, shape, and position of fins have not changed. The trigger (presumably recent) is behavioral. We shall wait to see what changes in form might still ensue.

As a final point, readers might acknowledge my argument, but dismiss the examples as trivial or peripheral. We all love flamingos, and *Cassiopea* might prick our interest (our bodies too, if we get in the way). Mochokids are amusing in aquaria. But can we view life upside down as any more than a funny little corner of natural history? All my examples are the dead-end adaptations of a few species; can turning upside down lead to anything fundamental and expansive?

As an important illustration from history (though almost surely an incorrect idea), life upside down once compelled attention as a leading speculation for the origin of vertebrates—the "worm that turned" theory, so to speak. Annelids and arthropods, the most complex of segmented invertebrates, develop ventral (bottom) nerve cords; the esophagus pierces the nerve cords and connects an even more ventral mouth to a central alimentary (gut) canal lying above the nerve cords. In vertebrates, the major nerve cord runs fore and aft in a dorsal (top) position, and the alimentary canal, including mouth and esophagus, lies entirely below.

These two designs seem quite incompatible and unrelated. But, and ironically in the context of my contrast between structural and functional views, the greatest of all structuralists, Geoffroy Saint-Hilaire himself, noted that an annelid turned on its back would look more than a bit like a vertebrate—for the ventral nerve cord would then become dorsal and lie above the alimentary canal. In solving one problem, others emerge: the mouth now opens atop the inverted worm. Geoffroy suggested, as an *ad hoc* solution straining credulity, that the old mouth and nerve-piercing esophagus simply disappeared, and that an entirely new opening (the vertebrate mouth) developed below the dorsal nerve cord, connecting directly with the gut canal, and no longer piercing the nervous system. (So many other differences plague the comparison—lack of any annelid structure resembling the notocord or gill slits of vertebrates, fundamental disparities in embryological development between the two groups, for example—that the worm theory never commanded general assent, though it remained a leading contender for nearly a century.)

Geoffroy never intended his comparison of vertebrate to inverted worm as an evolutionary speculation, but only as a structural comparison to buttress his remarkable theory that all animals shared a common architectural plan. (He also argued that the segments of an insect's external skeleton matched our internal vertebrae—and that insects literally lived within their own vertebrae. This comparison compelled the additional and astonishing conclusion, forthrightly maintained by Geoffroy, that insect legs are vertebrate ribs.)

Geoffroy also did not advance his comparison as a functional hypothesis about adaptation—he did not argue (as Lamarck might have done) that a worm's innovative behavior (in turning over) triggered an adaptive pressure for redesign. Quite the contrary. As a structuralist, he contended that belly and back are meaningless terms of human invention to describe a superficial orientation utterly without significance to what really matters—abstract structural laws of form and permitted pathways of change.

Today, we reject Geoffroy's speculation along with his approach to form and function. Life upside down affirms Lamarck's claim that substantial change in morphology usually arises as a consequence of behavioral triggers. The famous fourteenth-century motto of that upstart institution, New College, Oxford, seems to embody an essential truth about history as well as conduct: manners makyth man.

2 | Only His Wings Remained

THE CONVENTIONAL PROSE of twentieth-century science is lean and spare. But our Victorian predecessors delighted in leisurely detail, in keeping perhaps with the gingerbread on their houses and the shelves of bric-a-brac inside. Consider, for example, this extended (but most entertaining) description of sex and death in praying mantises, published by L.O. Howard in 1886:

A few days since, I brought a male of *Mantis carolina* to a friend who had been keeping a solitary female as a pet. Placing them in the same jar, the male, in alarm, endeavored to escape. In a few minutes, the female succeeded in grasping him. She first bit off his left front tarsus, and consumed the tibia and femur. Next she gnawed out his left eye. At this the male seemed to realize his proximity to one of the opposite sex, and began to make vain endeavors to mate. The female next ate up his right front leg, and then entirely decapitated him, devouring his head and gnawing into his thorax. Not until she had eaten all of his thorax except 3 millimeters did she stop to rest. All this while the male had continued his vain attempts to obtain entrance at the valvules, and he now succeeded, as she voluntarily spread the parts open, and union took place. She remained quiet for 4 hours, and the remnant of the male gave occasional signs of life by a movement of one of the remaining tarsi for 3 hours. The next morning she had entirely rid herself of her spouse, and nothing but his wings remained.

I cite this passage not merely for its style, but primarily for its substance—since it represents the first account I know of an all-time favorite among nature's curious facts. We have all heard that some animals can live after losing large portions of themselves, but we think of them as just scraping by in such a limited state, not as improving their skills. Our cliché about "running around like a chicken with its head cut off" underscores this reasonable assumption that reduced anatomy entails diminished competence. Yet male mantises, beheaded by a rapacious mate, not only continue their act of courtship and copulation but actually perform more persistently and successfully.

I want, as usual, to discuss the larger message behind this paramount oddity, but adequate treatment requires a long digression right back to Darwin himself. So bear with me, and we'll eventually get back to mantises and much more of what the biological literature calls "sexual cannibalism."

The *Descent of Man* is, without doubt, Darwin's most misunderstood book. Many people suppose that it represents Darwin's attempt to fit the facts of human evolution into his evolutionary perspective. But no direct facts existed when he published in 1871, for besides Neanderthal (a race of our own species, not an ancestor or any form of "missing link") no human fossils were discovered until the 1890s. Rather, the *Descent of Man* is an extended essay on the close biological relationship of humans with great apes and the possible modes of our physical and mental evolution from this common ancestry. But Darwin abhorred speculation; he never wrote a purely theoretical treatise. Even the *Origin of Species* is a compendium of facts pointing to a powerful conclusion. He would not have written a naked account of how it might have been, no matter how much he yearned to extend his evolutionary perspective to what he once called "the citadel itself"—the human mind.

The key to the *Descent of Man* is its situation as a relatively short preface to a large, two-volume work, *The Descent of Man and Selection in Relation to Sex.* Darwin could weave wonderful and extensive tapestries about central themes—so much so that his readers often lose the core in its extensive mantling. But all his books are solutions to specific puzzles; the rest, for all its brilliance, is superstructure. The coral reef book is about historical inference from contemporary results, the orchid book about imperfect adaptation based on parts available, the worm book about large effects accumulated by successive small changes (see essay 9 in *Hen's Teeth and Horse's Toes*). But because he loved detail, Darwin tells you more than you want to know about how insects fertilize orchids or how worms pull objects into their burrows—and you easily lose the kernel, the paradox, the gem of a problem that started the whole edifice.

The *Descent of Man* is a preface to such a problem. By 1871, twelve years after the *Origin of Species,* Darwin no longer needed to convince people of good will and mental flexibility that evolution had occurred; that battle had been won. But how does evolution work, what kind of world do we inhabit, and how can we know? Darwin's radical message lay in his claim that the beauties and harmony of nature are all byproducts of one primary process called natural selection: organisms struggle to achieve greater personal reproductive success—in modern parlance, to pass more of their genes into future generations (since they cannot preserve their bodies)—and that is all. No overarching laws about the good of species or ecosystems, no wise and watchful regulator in the skies—just organisms struggling.

But how can we know that the world is regulated by selection and not by some other evolutionary principle? Darwin's answer is brilliant, paradoxical, and usually misunderstood. Do not, he cautions, rest your case on what might seem to be the most elegant expression of selection—the beautiful, optimally designed adaptations of organisms to their environments: the aerodynamic perfection of a bird's wing or the streamlined beauty of a marlin. For good design is the expectation of most evolutionary theories (and of creationism as well, for that matter). There is nothing distinctively Darwinian about perfection. Instead, look for the oddities and imperfections that only occur if selection based on the reproductive success of individuals—and not on some other evolutionary mechanism—shapes the path of evolution.

The largest class of such oddities includes those structures and habits that plainly compromise the good design of organisms (and the ultimate success of species) but just as clearly increase the reproductive prowess of individuals bearing them. (My favorite examples are the tail feathers of peacocks and the huge, encumbering antlers of Irish elks, both adaptations in the struggle among males for access to, or acceptance by, females, but certainly not contributions to good design in the biomechanical sense.) Our world overflows with peculiar, otherwise senseless shapes and behaviors that function only to promote victory in the great game of mating and reproduction. No other world but Darwin's would fill nature with such curiosities that weaken species and hinder good design but bring success where it really matters in Darwin's universe alone—passing more genes to future generations.

Darwin realized that natural selection in its usual sense—increasing adaptation to changing local environments—would not explain this large class of features evolved to secure purely reproductive benefits for individuals. So he christened a parallel process, sexual selection, to explain this crucial evidence. He argued that sexual selection might work by combat among males or choice by females: the first to produce overblown weapons and instruments of display; the second to encourage those adornments and elaborate posturings that impel notice and acceptance (the nightingale does not sing for our delectation).

Humans enter the story at this point. Why did Darwin choose his long and detailed treatise on sexual selection as a home for his much shorter preface on the *Descent of Man*? The answer again lies in Darwin's fascination with specific puzzles and the contribution made by their solution to his larger goal. The *Descent of Man* has its anchor in a particular problem of human racial variation; it is not a waffling treatise on generalities. We can, Darwin argues, understand some racial differences, skin colors for example, as conventional adaptations to local environments (dark skin evolved several times independently and always in tropical climates). But surely we cannot argue that all the small, subtle differences among peoples—minor but consistent variations in the shape and form of noses and ears or the texture of hair—have their origin in what local environments ordain. It would be a vulgar caricature of natural selection to argue, by clever invention, that each insignificant nuance of design is really an optimal configuration for local circumstances (although many overzealous votaries continue to promote this view. A prominent evolutionist once seriously proposed to me that Slavic languages are full of consonants because mouths are best kept closed in cold weather, while Hawaiian has little but vowels because the salutary air of oceanic islands should be savored and imbibed). How then, if not by ordinary natural selection, did these small and subtle, but pervasive, racial differences originate?

Darwin proposes—and I suspect he was largely right—that different standards of beauty arise for capricious reasons among the various and formerly isolated groups of humans that people the far corners of our earth. These differences—a twist of the nose here, slimmer legs there, a curl in the hair somewhere else—are then accumulated and intensified by sexual selection, since those individuals accidentally endowed with favored features are more sought and therefore more successful in reproduction.

Look at the organization of the *Descent of Man* and you will see that this argument, not the generalities, provides its focus. The book begins with an overview of some 250 pages, all leading to a final chapter on human races and a presentation of the central paradox on the last page.

We have thus far been baffled in all our attempts to account for the differences between the races of man; but there remains one important agency, namely Sexual Selection, which appears to have acted powerfully on man, as on many other animals. . . . In order to treat this subject properly, I have found it necessary to pass the whole animal kingdom in review.

Darwin now has his handle for the real meat of his book, and he spends more than twice as much space, the next 500 pages, on a detailed account of sexual selection in group after group of organisms. Finally, in three closing chapters, he returns to human racial variation and completes his solution of the paradox by ascribing our differences primarily to sexual selection.

Sexual selection has sometimes been cast as a contrast or conflict with natural selection, but such an interpretation misunderstands Darwin's vision. Sexual selection is our most elegant confirmation of his central tenet that the struggle of individuals for reproductive success drives evolution—a notion that natural selection cannot adequately confirm because its products are also the predictions of other evolutionary theories (and also, for optimal design, of creationism itself). The proof that our world is Darwinian lies in the large set of adaptations arising *only* because they enhance reproductive success but otherwise both hinder organisms and harm species. Darwinian selection for reproductive success must be extraordinarily powerful if it can so often overwhelm other levels and modes of advantage.

We may now return to the blood meal of the mating mantis. W.H. Auden once wrote, with great understanding of our lives, that love and death are the only subjects worth the attention of literature. They are indeed the foci of Darwin's world, a universe of struggle for survival and continuity. But should they be conjoined? At first sight, nothing seems more absurd, less in keeping with any notion of order or advantage, than the sacrifice of life for a copulation. Should a male, in Darwin's world, not survive to mate again? Not necessarily, if he is destined for a short life and unlikely to mate again in any case, and if his "precious bodily fluids" (to cite the immortal line from *Dr. Strangelove*) will make a big difference in nourishing the eggs fertilized by his sperm within his erstwhile partner and current executioner.

After all, his body is so much Darwinian baggage. It cannot be passed to the next generation; his patrimony lies, quite literally, in the DNA of his sperm. Thus, sexual cannibalism should be a premier example of why we live in a Darwinian world—a classic curiosity, an apparent absurdity, made sensible by the proposition that evolution is fundamentally about struggle among organisms for genetic continuity. But how good is the evidence? (And now I must warn you—since this essay may be the most convoluted I have ever written—that this eminently reasonable argument for Darwinism has, in my assessment, very little going for it at present. Yet an alternative interpretation, for a different reason, affirms something even more fundamental about Darwinism and about the nature of history itself. Frankly, while I'm at the confessional, I began research for this essay on the assumption that such a lovely and reasonable argument for sexual selection would hold, and found myself quite surprised at the paucity of evidence. I also steadfastly refuse to avoid a subject because it is difficult. The world is not uncomplicated, and a restriction of general writing to the clear and uncontroversial gives a false view of how science operates and how our world works.)

A recent issue of the *American Naturalist,* one of America's three leading journals of evolutionary biology, featured an article by R.E. Buskirk, C. Frohlich, and K.G. Ross, "The Natural Selection of Sexual Cannibalism" (see bibliography). They develop a mathematical model to show that willing sacrifice of life to an impregnated partner will be to a male's Darwinian advantage if he can expect little subsequent success in mating and if the food value of his body will make a substantial difference to the successful development and rearing of his offspring. The model makes good sense, but nature will match it only if we can show that such males actively promote their own consumption. If they are trying like hell to escape after mating, and occasionally get caught and eaten by a rapacious female, then we cannot argue that sexual selection has directly promoted this strategy of ultimate sacrifice for genetic continuity.

Buskirk, Frohlich, and Ross are frank in stating that sexual cannibalism is not only rare in general but also much less common than other styles of consuming close relatives (as in sibling by sibling, or mothers by offspring; see essay 10 in *Ever Since Darwin* and essay 6 in *The Panda's Thumb*). Documented examples exist only for arthropods (insects and their kin), and only thirty species or so have been implicated (though the phenomenon may be quite common in spiders). They cite three examples as best cases.

1. The female praying mantis (*Mantis religiosa,* and several related species) will attack anything smaller than itself that moves. Since males are smaller than females in almost all insects, and since mating requires proximity, male mantises become a premier target. In his classic paper of 1935 (see bibliography), K. Roeder writes: "All accounts agree as to the ferocity of the female, and her tendency to capture and devour the male at any time, whether it be during the courtship or after copulation. . . . The female may seize and eat the male as she would any other insect."

   A male therefore approaches mating with the punch line of that terrible old joke about how porcupines do it: very carefully. He creeps up slowly, trying at all costs to keep out of the female's sight line. If the female turns in his direction, he freezes—for mantises ignore anything that doesn't move. Roeder writes: "So extreme is this immobility that if a male is in the act of raising a leg when first the female is detected, it will be kept poised in the air for some time, and many curious positions may be observed." Thus, the male continues to approach like a child playing the street game of "red light"—drawing near while his adversary and potential mate averts her eyes, freezing instantly when she looks around (although the penalty for apprehended motion is death, not a return to the starting line). If the male succeeds in creeping up within springing distance, he makes a fateful leap to the female's back. If he misses, he's mantis fodder; if he succeeds, he achieves the Darwinian *summum bonum* of potential representation in the next generation. After mating, he falls off as far away as he can and then skedaddles with dispatch.

So far, the story sounds little like a tale of active male conspiracy in his own demise—the requirement, please remember, for an argument that males are directly selected for sexual cannibalism. Perhaps males are simply trying their darndest to get away, but don't always make it. The strong argument inheres in that great curiosity mentioned at the outset of this essay: decapitated males perform *better* sexually than their intact brothers. Roeder has even discovered the neurological basis for this peculiar situation. Much of insect behavior is "hard wired," so unlike the flexibility of our own actions (and a primary reason why sociobiological models for ants work so poorly for humans). Copulatory movements are controlled by nerves in the last abdominal ganglion (near the back end). Since it would be inconsistent with normal function (and unseemly as well) for males to perform these copulatory motions continually, they are suppressed by inhibitory centers located in the subesophageal ganglion (near the head). When a female eats her mate's head, she ingests the subesophageal ganglion, and nothing remains to inhibit copulatory movements. What remains of the male now operates as a nonstop mating machine. It will try to mount anything—pencils, for example—of even vaguely appropriate size or shape. Often it finds the female and succeeds in making of its coming death the Darwinian antithesis of what Socrates called "a state of nothingness."

2. A hungry female black widow spider is also a formidable eating machine, and courting males must exercise great circumspection. On entering a female's web, the male taps and tweaks some of her silk lines. If the female charges, the male either beats a hasty retreat or sails quickly away on his own gossamer. If the female does not respond, the male approaches slowly and cautiously, finally cutting the female's web at several strategic points, thereby reducing her routes of escape or attack. The male often throws several lines of silk about the

female, called, inevitably I suppose, the "bridal veil."
They are not strong, and the larger female could surely
break them, but she generally does not, and copulation,
as they like to say in the technical literature, "then en-
sues." The male, blessed with paired organs for transferring sperm, inserts one palp, then, if not yet attacked by
the female, the other. Hungry females may then gobble
up their mates, completing the double-entendre of a
consummation devoutly to be wished.

The argument for direct selection of sexual cannibalism rests upon two intriguing phenomena of courtship.
First, the tip of the male's palp usually breaks off during
copulation and remains behind in the female. Males,
thus rendered incomplete, may not be able to mate
again; if so, they have become Darwinian ciphers, ripe for
removal. (An interesting speculation identifies this broken tip as a "mating plug" selected to prevent the entry
of any subsequent male's sperm. Such natural *post factum*
chastity belts are common, and of diverse construction,
in the world of insects and would make a fine subject for
a future essay on the same issue of why sexual selection
identifies our evolutionary world as Darwinian.) Second,
males show far less avidity and caution in scramming
after the fact than they did in approaching before. K.
Ross and R.L. Smith write (see bibliography): "Males
that succeeded in insemination lingered in the vicinity of
their mates or wandered leisurely away. This was in
marked contrast with the initial cautious approach and
escape strategies characteristic of males prior to insemination."

3. Females of the desert scorpion *Paruroctonus mesaensis* are
extremely rapacious and will eat anything small enough
that they can detect. "Any moving object in the proper
size range is attacked without apparent discrimination"
(G.A. Polis and R.D. Farley, see bibliography). Since
males are smaller than females, they become prime targets and are consumed with avidity. This indiscriminate
rapacity presents quite a problem for mating, which, as
usual, requires some spatial intimacy. Males have there-

fore evolved an elaborate courtship ritual, in part to suppress the female's ordinary appetite.

The male initiates a series of grasping and kneading movements with his chelicerae (minor claws), then grabs the female's chela (major claw) with his own and performs the celebrated *promenade à deux,* a reciprocal and symmetrical "dance," pretty as anything you'll see at Arthur Murray's. These scorpions do not inseminate females directly by inserting a penis, but rather deposit a spermatophore (a packet of sperm) that the female must then place into her body. Thus, the male leads the female in the *promenade* until he finds an appropriate spot. He deposits the spermatophore, usually on a stick or twig, then bats or even stings her, disengages, and runs for his life. If good fortune smiles, the female will let him go and pay proper attention to inserting his spermatophore. But, in two cases out of more than twenty, Polis and Farley found the female munching away on her mate while his spermatophore remained on a nearby stick, presumably for later ingestion through a different aperture.

What evidence, then, do these cases provide for selection of sexual cannibalism among males? Do males, for the sake of their genetic continuity, actively elicit (or even passively submit to) the care and feeding of their fertilized eggs with their own bodies? I find little persuasive evidence for such a phenomenon in these cases, and I wonder if it exists at all —although the argument would provide an excellent illustration of a curiosity that makes little sense unless the evolutionary world works for reproductive success of individuals, as Darwinism argues.

The scorpion story, despite its citation among best cases, provides no evidence at all. As I read Polis and Farley, I note only that males try their best to escape after copulation and succeed in a great majority of cases (only two failed). Indeed, their mating behavior, both before and after, seems designed to avoid destruction, not to court it. Before, they turn off the female's aggressive instincts by marching and stroking. After, they hit and run. That a few fail and get eaten reflects the inevitable odds of any dangerous game that must be played.

Black widow spiders and praying mantises offer more to the theory of direct selection for destruction among males. The spiders seem to be as cautious as scorpions before, but quite lackadaisical after, making little if any attempt to escape from the female's web. In addition, if the mating plug that they leave in the female debars them from any future patrimony, then they have fully served their Darwinian purpose. As for mantises, the better performance of a headless male might indicate that sex and death have been actively conjoined by selection. Yet, in both these cases, other observations render more than a bit ambiguous any evidence for active selection on males.

As a major problem for both mantises and spiders, we have no good evidence about the frequency of sexual cannibalism. If it occurred always or even often and if the male clearly stopped and just let it happen, then I would be satisfied that this reasonable phenomenon exists. But if it occurs rarely and represents a simple failure to escape, rather than an active offering, then it is a byproduct of other phenomena, not a selected trait in itself. I can find no quantitative data on the percentage of eating after mating either in nature or even in the more unsatisfactory and artificial conditions of a laboratory.

For mantises, I find no evidence for the male's complicity in his demise. Males are cautious beforehand and zealous to escape thereafter. But the female is big and rapacious; she makes no distinction between a smaller mantis and any other moving prey. As for the curious fact of better performance in decapitated males, I simply don't know. It could be a direct adaptation for combining sex with consumption, but other interpretations fare just as well in our absence of evidence. Hard-wired behavior must be programmed in some way. Perhaps the system of inhibition by a ganglion in the head and activation by one near the tail evolved in an ancestral lineage long before sexual cannibalism ever arose among mantises. Perhaps it was already in place when female mantises evolved their indiscriminate rapacity. It would then be co-opted, not actively selected, for its useful role in sexual cannibalism. After all, the same system works for females too, although their behavior serves no known evolutionary function. Decapitate a female mantis and you also unleash sexual behavior, including egg laying. If one wishes to argue that the system must have been actively evolved because the female tends to eat first just that portion of the male that unleashes sexuality, I reply with a bit of biology at its most basic: heads are in front and females encounter them first as the male approaches.

The black widow story is also shaky. Males may not try to escape after mating, but is this an active adaptation for consumption or an automatic response to the real adaptation—breaking of the sexual organ and deposition of a mating plug in the female (for such an injury might weaken the male and explain his subsequent lassitude)? Also, male black widows are tiny compared with their mates—only 2 percent or so of the female's weight. Will such a small meal make enough of a difference? Finally, and most importantly, how often does the female partake of this available meal? If she always ate the exhausted male after mating, I would be more persuaded. But some studies indicate that sexual cannibalism may be rare, even though clearly available as an option for females. Curiously, several articles report that males often stay on the female's web until they die, often for two weeks or more, and that females leave them alone. Ross and Smith, for example, noticed only one case of sexual cannibalism and wrote: "Only one male of those we observed to succeed in inseminating a female was eaten by its mate immediately after mating. However, several were later found dead in their mates' webs."

Why then, in this disturbing absence of evidence, does our literature abound with comments on the obvious evolutionary good sense of sexual cannibalism? For example: "Under some conditions selection should favor the consumption of males by their mates. His probability of being cannibalized should be directly proportional to the male's future expectation of reproduction." Or, "Successful males would best serve their biological interests by presenting themselves to their mates as a post-nuptial meal."

In this hiatus between reasonable hope and actual evidence, we come face to face with a common bias of modern Darwinism. Darwinian theory is fundamentally about natural selection. I do not challenge this emphasis, but believe that we have become overzealous about the power and range of selection by trying to attribute every significant form and behavior to its direct action. In this Darwinian game, no prize is sweeter than a successful selectionist interpretation for phenomena that strike our intuition as senseless. How could a male become a blood meal after mating if selection rules our world? Because, in certain situations, he increases his own reproductive success thereby, our devoted selectionist responds.

But another overarching, yet often forgotten, evolutionary principle usually intervenes and prevents any optimal match between organism and immediate environment—the curious, tortuous, constraining pathways of history. Organisms are not putty before a molding environment or billiard balls before the pool cue of natural selection. Their inherited forms and behaviors constrain and push back; they cannot be quickly transformed to new optimality every time the environment alters.

Every adaptive change brings scores of consequences in its wake, some luckily co-opted for later advantage, others not. Some large females evolve indiscriminate rapacity for their own reasons, and some males suffer the consequences despite their own evolutionary race to escape. Designs evolved for one reason (or no reason) have other consequences, some fortuitously useful. Male mantises can become headless wonders; male black widows remain on the female's web. Both behaviors may be useful, but we have no evidence that either arose by active selection for male sacrifice. Sexual cannibalism with active male complicity should be favored in many groups (for the conditions of limited opportunity after mating and useful fodder are often met), but it has evolved rarely, if ever. Ask why we don't see it where it should occur; don't simply marvel about the wisdom of selection in a few possible cases. History often precludes useful opportunity; you cannot always get there from here. Females may not be sufficiently rapacious, or they may be smaller than males, or so limited in behavioral flexibility that they cannot evolve a system to turn off a general inhibition against cannibalism only after mating and only toward a male.

Our world is not an optimal place, fine tuned by omnipotent forces of selection. It is a quirky mass of imperfections, working well enough (often admirably); a jury-rigged set of adaptations built of curious parts made available by past histories in different contexts. Darwin, who was a keen student of history, not just a devotee of selection, understood this principle as the primary proof of evolution itself. A world optimally adapted to current environments is a world without history, and a world without history might have been created as we find it. History matters; it confounds perfection and proves that current life transformed its own past. In his famous disquisition on the ages of man—"All the world's a stage"—Jaques, in *As You Like It,* speaks of "this strange eventful history." Respect the past and inform the present.

Postscript

In the light of my ever-growing doubts about the existence of sexual cannibalism (despite its plausibility in theory)—as prominently displayed in the personal odyssey of the essay itself—I was delighted by a report from the 1984 annual meeting of the Society for Neuroscience. E. Liske of West Germany and W.J. Davis of the University of California at Santa Cruz videotaped and analyzed courtship behavior for dozens of matings in Chinese praying mantises. No female ever decapitated or ate a male. Instead, frame-by-frame analysis revealed a complex series of behaviors, seemingly directed (at least in part) towards suppressing the natural rapacity of famales. Male behavior includes visual fixation, antennal oscillation, slow approach, repetitive flexing of the abdomen, and a final flying leap towards the female's back. Liske and Davis suggest that previous reports of decapitation may represent aberrant behavior of captive specimens (though sexual cannibalism may still be normal behavior in strains or species other than those studied by Liske and Davis. There is no such thing as *the* praying mantis, given nature's propensity for diversity.) In any case, I am even more persuaded that sexual cannibalism is a phenomenon without proven examples, and that the reasons for its rarity (or non-existence) form a far more interesting subject (and an appropriate shift of emphasis) than the one that first inspired my research for the essay—reasons for the presumed (and now dubious) existence itself.

I often argue that the best test for legends is the extent of their seepage into popular culture. In *Sherlock Holmes and the Spider Woman* (1944), one of the innumerable, yet wonderful, Rathbone-Bruce anachronisms that pit Holmes against Hitler and assorted enemies, Holmes unmasks an entomologist *poseur* (and murderer of the true scientist) by catching several subtle fallacies in his speech. The phony calls terraria "glass cages," but then really gives himself away when he says of black widow spiders: "They eat their mates, I'm told." Holmes responds: "You said you were told the black widows eat their mates. Any scientist would know it." I shall be waiting for the next update (who is playing Charlie Chan these days?).

3 | Sex and Size

AS AN EIGHT-YEAR-OLD COLLECTOR of shells at Rockaway Beach, I took a functional but non-Linnaean approach to taxonomy, dividing my booty into "regular," "unusual," and "extraordinary." My favorite was the common slipper limpet, although it resided in the realm of the regular by virtue of its ubiquity. I loved its range of shapes and colors, and the pocket underneath that served as a protective home for the animal. My appeal turned to fascination a few years later, when I both entered puberty and studied some Linnaean taxonomy at the same time. I learned its proper name, *Crepidula fornicata*—a sure spur to curiosity. Since Linnaeus himself had christened this particular species, I marveled at the unbridled libido of taxonomy's father.

When I learned about the habits of *C. fornicata*, I felt confident that I had found the key to its curious name. For the slipper limpet forms stacks, smaller piled atop larger, often reaching a dozen shells or more. The smaller animals on top are invariably male, the larger supporters underneath always female. And lest you suspect that the topmost males might be restricted to a life of obligate homosexuality by virtue of their separation from the first large female, fear not. The male's penis is longer by far than its entire body and can easily slip around a few males to reach the females. *Crepidula fornicata* indeed; a sexy congeries.

Then, to complete the disappointing story, I discovered

56

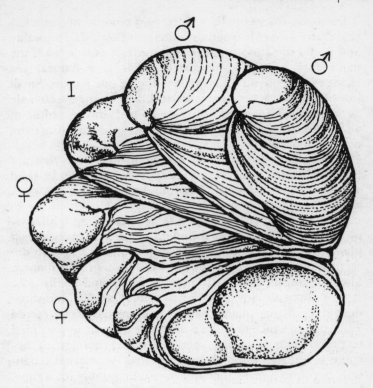

A *Crepidula* stack, with sexes identified. Bottom members are female; the smaller individuals on top are male. The animal in the middle (labeled I) is undergoing a transition from male to female. REPRINTED FROM NATURAL HISTORY.

that the name had nothing to do with sex. Linnaeus had described the species from single specimens in museum drawers; he knew nothing of their peculiar stacking behavior. *Fornix* means "arch" in Latin, and Linnaeus chose his name to recognize the shell's smoothly domed shape.*

---

*I have since learned that the story is not so disappointing and one-directional (ostensible sense to coincidence) as I had concluded when I wrote the essay. Linguistic history has provided a back formation from

Disappointment finally yielded to renewed interest a few years later when I learned the details of *Crepidula*'s sexuality and found the story more intriguing than ever, even if the name had been a come-on. *Crepidula* is a natural sex changer, a sequential hermaphrodite in our jargon. Small juveniles mature first as males and later change to female as they grow larger. Intermediate animals in the middle of a *Crepidula* stack are usually in the process of changing from male to female.

The system works neatly for all involved. *C. fornicata* tends to live in relatively muddy areas but must find a solid substrate for attachment. The founding member of a stack affixes to a rock or an old shell. Elaine Hoagland, in an exhaustive study of *Crepidula*'s sex changes (see bibliography), observed that these founders can then actively attract planktonic larvae as they metamorphose and begin to descend—presumably by some chemical lure, or pheromone. She set out six pots with suitable rocky and shelly substrates: three already occupied by adult *Crepidula* and three devoid of living snails. Pots containing adults attracted 722 young, while only 232 descended upon unoccupied territory. The founding member grows quickly to become a female, while the young spat on top automatically becomes a male. The union remains stable for a time, but eventually the male grows up and turns into a female. The pair of females can then attract other small *Crepidulas,* which become well-supplied males. The stack grows, always maintaining an ample number and ratio of males and females.

This curious system provides a particularly interesting example of a general phenomenon in nature. Sex change might go either way (or both) during growth, from male to female or from female to male. Both phenomena occur, but *Crepidula*'s pattern of male first and female later, called *prot-*

---

morphological arches to sex. In *A Browser's Dictionary* (Harper and Row, 1980), John Ciardi reports (p. 137): ". . . because the Romans used . . . arched brickwork in the underground parts of great buildings, and because the poor and the prostitutes of Rome lived in such undergrounds . . . early Christian writers evolved the verb *fornicari,* to frequent brothels. The whores of Pompeii worked in similar stone cribs."

andry (or male first) is by far the more common. (Creatures that are first female and then male are *protogynous,* or female first.) Protandry seems to represent the prevalent path of changing sex, with protogyny as a rarer phenomenon evolved under special (but not particularly uncommon) circumstances. Why should this be?

The answer pricks one of our old prejudices and false extrapolations to all of nature from the animals we know best, ourselves and other mammals. We think of males as large and powerful, females as smaller and weaker, but the opposite pattern prevails throughout nature—males are generally smaller than females, and for good reason, humans and most other mammals notwithstanding. Sperm is small and cheap, easily manufactured in large quantities by little creatures. A sperm cell is little more than a nucleus of naked DNA with a delivery system. Eggs, on the other hand, must be larger, for they provide the cytoplasm (all the rest of the cell) with mitochondria (or energy factories), chloroplasts (for photosynthesizers), and all other parts that a zygote needs to begin the process of embryonic growth. In addition, eggs generally supply the initial nutriment, or food for the developing embryo. Finally, females usually perform the tasks of primary care, either retaining the eggs within their bodies for a time or guarding them after they are laid. For all these reasons, females are larger than males in most species of animals.

This system can be overridden when males evolve a form of competition with other males that favors large size for success in gaining sexual contact with females. Such forms of competition are wasteful in terms of such theoretical concepts as "the good of the species." But Darwinism is about the struggle of individual organisms to pass more of their genes to future generations. The best indication that our world is Darwinian lies with these cases of evolution for individual advantage alone—as when males become larger because they compete as individuals in battle or sexual display for access to females.

Competition of this form generally requires a fair degree of intelligence, since such complex actions imply flexible and extensive behavioral repertoires. Thus, we tend to find the unusual or reversed pattern of larger males in so-called higher creatures of substantial brain. This correlation of complexity and mental power probably explains why, of all groups with a large number of sequential hermaphrodites, only vertebrates have evolved protogyny as a more common pattern than protandry. When we look at the natural history of most protogynous fish, we see that behavioral imperatives based on male-male competition have conditioned the pattern of females first, changing to larger males. Douglas Y. Shapiro, for example, studied sex reversal in *Anthias squamipinnis,* a shallow-water tropical marine fish that lives among coral reefs in stable social groups averaging eight adult females to one male (see bibliography). Males may compete intensely to retain and maintain their groups. The removal of a male induces a female to change sex, and this transition includes a set of features all conducive to keeping charge of several females: change to more gaudy color, longer fin spines, more elaborate caudal fin streamers, and larger size.

The distribution of protandry and protogyny provides an even better illustration of nature's preference for larger females than the simple documentation of permanently smaller males in insects or angler fishes. Permanent males and females represent static systems that may maintain their relationships of size for a set of other reasons. But when we find that *active* change of sex usually proceeds from male to female, we must seek some direct reason rooted in the general advantages of larger female size.

We might seek a still better illustration, one, unfortunately, that animals, as a constraint of their mode of growth, will not be able to provide. Ideally, we would like to find a creature that changes sex in either direction but becomes female when it grows larger and male when it becomes smaller. Can we hope for such an ideal case in nature, a total confirmation of a general principle all wrapped up in a single creature? (As long as we must wrap the principle in several creatures, we shall be haunted by the distressing possibility that we have it all wrong—that protogyny dominates in fishes, not because they are advanced behaviorally and illustrate Darwin's principle of individual competition, but as a consequence of some unknown and peculiar property of fishness. If, however, we can find both phenomena in the same creature, a unified explanation seems assured.) But do we have a right to expect such an ideal example from nature? Animals, after all, with very rare exceptions, never grow smaller and will therefore not serve. One of the earliest articles on sex change in *Crepidula,* written in 1935, ended with these words: "Sexual transformation in *Crepidula,* like metamorphosis in other animals, can be hastened or retarded experimentally, but it cannot be reversed."

Nature has come through again—she always does. The ideal organism has surfaced. It is a plant, the general subject, unfortunately, of my woeful and abysmal ignorance. Plants can undergo substantial reduction in size, for several reasons and without expiring. Our example is a common and attractive inhabitant of local eastern woodlands, *Arisaema triphyllum,* the jack-in-the-pulpit. Results were recently reported by my friend David Policansky in the staid Proceedings of the National Academy of Sciences (see bibliography). (I confess that my previous attention to this plant was virtually confined to wondering whether its plural form included one jack and several pulpits, as in most words, or several jacks and one pulpit, as in those old bugbears of high school grammar, attorneys-general and mothers-in-law. I note that this matter must confuse others as well, because the two references I have found to Policansky's work both studiously avoid the issue and, in defiance of the rules of grammar, use the singular in all cases. I will opt for several pulpits, even though I know that each one carries a jack. Or are they like sheep after all?)*

The flowers of most (but by no means all) plants contain both male and female structures. But jack-in-the-pulpits are either one or the other. The sexual part of the blossom

---

*One reader suggested the obvious and elegant solution to this dilemma of the ages—jacks-in-pulpits. How incredibly stupid of me not to have thought of it myself.

contains either anthers, the male's sexual structure, or ovaries capped with stigmas. Smaller plants, the males, have one leaf, while the larger females usually grow two leaves. During a three-year study at Estabrook Woods in Concord, Massachusetts, Policansky marked and recorded 2,038 jack-in-the-pulpits; 1,224 were male with a mean height of 336 mm, while 814 females averaged 411 mm in height.

The so-called "size advantage" model of sex change predicts that, for the usual case of smaller males, a transition from male to female should occur where any further increase in size begins to benefit a female (in terms of seeds that can be produced) more than a male. (Remember that small males can produce a superabundance of sperm, and larger size therefore offers relatively little additional advantage, while the benefit to females can be substantial.) Citing data on the increase in sperm and seed number with size, Policansky calculated that, in theory, this transition in jack-in-the-pulpits should occur at a height of 398 mm. He then found that, in nature (or, at least, in Concord), 380 mm is the watershed—a very close agreement with theory. Below this height, he found more males than females; above, more females than males.

He was also able to ascertain directly that male plants tended to change to female as they grew larger in the normal course of life. Moreover, and this is the key observation, individuals changed from female to male in the more unusual circumstances that occasionally lead a plant to become smaller. Size decrease occurred for three reasons: when part of the plant was eaten (if Jill breaks her crown, Jack comes after); when the plant became shaded and, consequently, stunted in growth; and when it had set an unusually large number of seeds the season before, thus also inhibiting growth in size by diverting most energy to the seeds themselves.

Thus, with change in both directions conforming to the size advantage model and following nature's usual pattern of smaller males and larger females, the jack-in-the-pulpit provides, all by itself, a lovely illustration of the errors in our usual, narrow perceptions and assumptions about the relative size of sexes—and an excellent confirmation of an important principle in Darwinian biology. It will also help us to understand why, if *man* is truly the measure of all things, Jill will need an enlarged pulpit.

4 | Living with Connections

LA GRANDE GALERIE of the Muséum d'histoire naturelle in Paris has been closed for fifteen years. This great space, supported by iron and roofed in glass, is no longer structurally sound. Like the capacious nineteenth-century railroad stations that served as its model, La Grande Galerie has passed into history. Its exhibits, too, reflect the thoughts and concerns of another age, the expansive and aggressive Victorian era that took so seriously, as a guide for collection and display, the words of Genesis (1:22): "Be fruitful and multiply, and fill the waters in the seas, and let fowl multiply in the earth." If modern museums emphasize intimacy, good lighting, tasteful display, and well-chosen words, their Victorian predecessors judged quality by quantity and crammed as many large animals as possible into their vast open spaces. At Lord Rothschild's museum in Tring, the stuffed zebras are supine, so that several tiers may be fitted from floor to ceiling.

La Grande Galerie is the granddaddy of this superseded style. Built in 1889, and unchanged since, its skeletons and stuffed animals occupy every available inch. The great central pyramid almost reaches the high glass ceiling. One side is all zebras, another all antelopes; six giraffes crown the summit. The dust is thick, the hall dark and empty; eerie silence marks a dingy majesty.

The companion hall, La galerie d'anatomie comparée, is smaller, well lit, and still open. Its style is identical—row upon endless row, tier upon tier of blanched skeletons. I wandered up and down the aisles, marveling at a long row of walruses and five superposed tiers of monkey skulls. Then I passed by cabinet 106 and stopped short. It contains a sideshow to offset the neighboring forest of sleek lions, and to remind complacent Victorians that nature can be capricious and cruel, as well as bountiful. Cabinet 106 holds a collection of teratological specimens, skeletons of deformed and abnormal births. Most are human and most represent that puzzling and frightening phenomenon of joined birth, or "Siamese" twinning. Skeleton A8597 has two heads, three arms, and two legs; A8613 has four arms, two legs, and two heads projecting from the ends of a joined vertebral column; A8572 is almost normal, but a tiny, headless brother with arms and legs projects from his chest. All are small and clearly died at birth or soon thereafter.

One skeleton stands out for its considerably larger size. A8599 is (or are)—and this is the issue we shall soon discuss —twin girls with two well-formed heads and upper bodies with four full arms. Two distinct vertebral columns nearly join at their base, and only two well-formed legs extend below. The label reads *monstre humain dicéphale,* or "two-headed human monster." But A8599 was born live and survived several months. The twins were baptized and given names. The label records this poignant detail and includes, under the number and description, the simple identification "Ritta-Christina."

I mused much over Ritta and Christina, wondering about their life and death. Yet I would not have made the transition from troubled thought to essay had I not discovered, quite by accident, a dusty old tome in a bookstore two days later—volume 11, for 1833, of the *Memoirs of the Royal Academy of Sciences.* It contained a long monograph by the great medical anatomist Etienne Serres: *Théorie des formations et des déformations organiques, appliquée à l'anatomie de Ritta Christina, et de la duplicité monstrueuse* ("Theory of organic development and deformation, applied to the anatomy of Ritta Christina, and to duplicate monsters in general").

Anyone who does not grasp the close juxtaposition of the

vulgar and the scholarly has either too refined or too compartmentalized a view of life. Abstract and visceral fascination are equally valid and not so far apart. Two days before, I had seen young schoolchildren standing before Ritta-Christina in open-mouthed awe or horror, soon masked by forced humor. Now I learned that France's finest medical anatomist had dissected Ritta-Christina and used her to support a general theory of organic (not only human) embryology. Both themes seemed equally compelling to me; indeed, I had wallowed in both myself for two days. The children might not have generalized, but I have no doubt that M. Serres once gulped, as well as thought. I bought the book.

Ritta and Christina were born on March 23, 1829, to poor parents in Sardinia. Times were hard and social mobility scarcely possible in ordinary circumstances. Parents today would receive pity and experience only sorrow; in 1829, realistic people, whatever their private feelings, must have recognized that such a child represented potential and substantial revenue, otherwise quite unobtainable.* Thus, the parents of Ritta-Christina scraped together some funds and brought her to Paris, hoping to display her at fancy prices. The Hottentot Venus had provoked enough protest fifteen years earlier (see essay 19); but she was whole, however exotic. Public sensibilities had limits, and the authorities forbade any open display of Ritta and Christina. But she was shown privately, many times too often—for she died, in part from overexposure, after five months of life.

I have consciously switched back and forth from singular to plural in describing Ritta-Christina. When the vulgar and scholarly meet, a common question often underlies our

---

*I am tempted to revise this sentence, and assert a universality transcending time, in the light of two back-to-back stories from the *New York Times* of November 23, 1984—first, that the heirs of Barney Clark (the deceased first recipient of a mechanical heart) have filed a $2 million lawsuit against *Reader's Digest* for breaking their contract to publish a book on the case by Mr. Clark's widow; second, that Baby Fae's parents (she of the baboon heart transplant) have sold exclusive rights for their story to *People Magazine*.

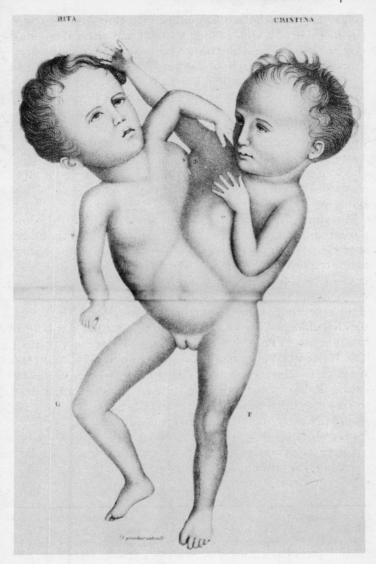

Ritta-Christina, drawn from life. FROM SERRES, 1833.

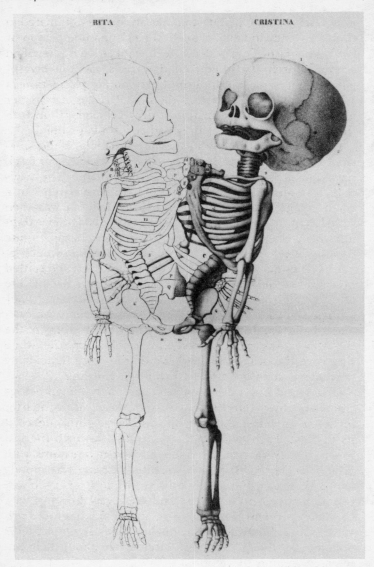

RITA   CRISTINA

The skeleton of Ritta-Christina after Serres's preparation and still on display in Paris. FROM SERRES, 1833.

joint fascination. One question has always predominated in this case—individuality. Was Ritta-Christina one person or two? This issue inspired the feeble jokes of my terrified schoolchildren. It also motivated Serres's scientific investigation. The same question underlay public fascination in 1829. When Ritta-Christina died, a Parisian newspaper wrote: "Already it is a matter of grave consideration with the spiritualists, whether they had two souls or one."

One or two? Through all scholarly excursions and sideshow huckstering, this single question has focused our fascination since Siamese twinning received its name. The originals, Eng and Chang, were born of Chinese parents in a small village near Bangkok in 1811 (Thailand was then called Siam). During the late 1820s and 1830s, they exhibited themselves in Europe and America and became quite wealthy. They decided to live in North Carolina where, at age 44, they married two sisters of English birth and settled down in two neighboring households to a comfortable life as successful (and, yes, even slaveholding) farmers. They switched houses at three-day intervals, traveling the one-and-a-half-mile distance by carriage. By the customs of the day, Chang was unquestioned master in his domicile, while Eng gave the orders *chez lui*. The unions were undeniably productive, for Chang had 10 children and Eng 12.

Chang and Eng were physically complete human beings connected by a thin band of tissue, three and a quarter inches at its widest and only one and five-eighths inches at its thickest. Each had a full set of parts down to the last toenail. They carried on independent conversations with visitors and had distinct personalities. Chang was moody and melancholy and finally took to drink; Eng was quiet, contemplative, and more cheerful. Yet even they, history's most independent Siamese twins, apparently harbored private doubts about their individuality. They signed all legal documents "Chang Eng" and often spoke about their ambiguous feelings of autonomy.

But what of Ritta and Christina, whose bodily independence did not extend below the navel? They seemed, at first glance, to be two people above and only one below. The old

Eng and Chang, the original Siamese twins. PHOTO COURTESY OF
THE GRANGER COLLECTION.

cultural criterion of head and brain might have suggested
an easy resolution—two heads, two people. But as a scientist, Serres resisted this facile answer, for he had also studied Siamese twins with one head, two arms, and four legs.
He reasoned that a uniformity of process must underlie
both types of twinning and could not accept the simplistic

resolution—one person if zipped halfway but starting from the top; two if zipped from the bottom.

Serres struggled with this momentous issue for 300 pages and finally concluded that Ritta *and* Christina were two people. His arguments and basic style of science belong to another era in the history of biology. They are worth recounting if only because few intellectual exercises can be more rewarding than an examination of how radically different systems of thought treat a common subject of mutual interest. I also believe that Serres was at least half wrong.

Serres represented the great early nineteenth-century tradition of romantic biology, called *Naturphilosophie* ("nature philosophy") in Germany and transcendental morphology in his native France. If modern morphologists study form either to determine evolutionary relationships or to discover adaptive significance by examining function and behavior, Serres and his colleagues pursued markedly different goals. They were obsessed with the idea that some overarching, transcendental law must underlie and regulate all the apparent diversity of life.

These laws, in the Platonic tradition, must exist before any organism arises to obey their regularities. Organisms are accidental incarnations of the moment; the simple, regulating laws reflect timeless principles of universal order. Biology, as its primary task, must search for underlying patterns amidst the confusing diversity of life. In short, biologists must seek the "laws of form."

Serres contributed to the transcendental tradition by extending its concerns to embryology. Most of his colleagues had emphasized the static form of adults by searching for underlying patterns in final products alone. But organisms grow their own complexity from egg to adult. If laws of form regulate morphology, then we must discover principles for dynamic construction, not merely for relationships among finished creatures.

Serres's monograph on Ritta-Christina begins with an abstruse 200-page dissertation on the principles of morphology and their application to embryology. Unless we sneak a peek at the alluring plates in the back (including the

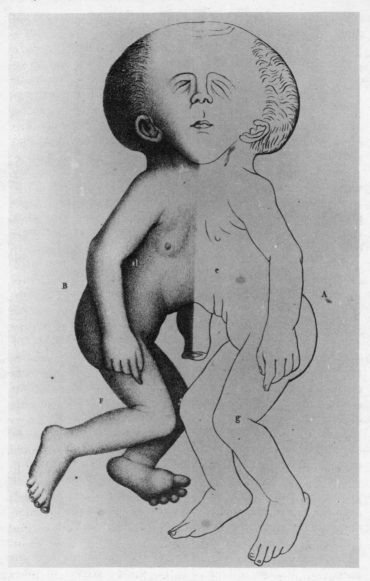

A Siamese twin pair of conformation opposite to Ritta-Christina: one upper and two lower bodies. FROM SERRES, 1833.

three figures reproduced with this essay), we hear nothing of the famous Sardinian twins until our senses are numbed by generality. This organization, in itself, reflects a style of science strikingly different from our own. We maintain an empirical perspective and like to argue that generalities arise from the careful study and collation of particulars. Any modern embryologist would discuss Ritta-Christina first and only venture some short and cautious conclusions at the end. But Serres, as a transcendentalist, believed that laws of form existed before the animals that obeyed them. If abstraction preceded actuality in nature, why not in human creativity as well? Thought and theory first, application later. (Neither extreme well represents the intricate interplay of fact and theory that regulates our actual practice of science. Still, I suspect that Serres's "inverted" order is no worse a distortion of complex reality than our modern stylistic preferences.)

In the first pages of his monograph, Serres tries to reduce the embryology of all animals to three basic laws of "organology." First, by the law of *eccentric development*, known otherwise as the law of circumference to center, organs form initially at the edge of the developing embryo and then migrate toward the center. Second, by the law of *symmetry*, organs that become single and central in adults begin as double symmetrical rudiments on opposite edges of the developing embryo. Third, by the law of *affinity*, these symmetrical rudiments are drawn towards each other until they fuse in the center to form a single adult organ. (Let me be charitable and simply state that these laws are unwarranted extensions of patterns that operate occasionally in development. Serres was writing before the establishment of cell theory and just a few years after Karl Ernst von Baer's discovery of the human ovum. His formal approach to morphology, so foreign to a world that can assess cellular and even molecular causes, fit the knowledge and mores of his own era.)

Two hundred pages later, when Serres finally discusses his dissection of Ritta-Christina, we understand why he devoted so much preceding space to the three primary laws of organology—for they provide his solution to the great dilemma of individuality. Ritta and Christina are two people, albeit imperfect, and the laws of form proclaim *their* status.

No one quarreled with the double verdict on Ritta and Christina from the waist up; the dilemma had always centered upon their well-formed, but clearly single, lower half —one anus, one vaginal opening, two legs. If she were two people all the way from stem to stern, how could her lower half form so well in the shape of one? How could the incomplete parts of two separate creatures fuse and blend into a form indistinguishable from the lower half of such unambiguous singletons as you or me?

Serres used his laws of organology to render Ritta and Christina's lower half as the conjoined product of two people. After all, the harmonious, well-formed single organs of ordinary individuals arise (by the law of symmetry) as separate and double parts at the embryo's edge, and then move inward (by the law of circumference to center), eventually meeting and fusing (by the law of affinity) into one integral organ. If our single heart, stomach, and liver begin as two symmetrical rudiments (actually, they don't, but Serres thought they did), then why should we view the presence of a single, well-formed organ in Ritta and Christina's lower half as any argument against its construction from the mingled and melded parts of two embryonic individuals? If the twins have but one uterus, then the right half came from Ritta, the left from Christina. The two rudiments formed at the embryonic edges, in regions unambiguously assigned to Ritta or to Christina (law of symmetry). They moved toward the midline (circumference to center) and joined there (law of affinity) to form a single organ.

Serres announced proudly that his laws of form had resolved the great dilemma in favor of duality: "How could we possibly have conceived that each child furnished half of an organ common to both, if the law of eccentric development had not taught us that single organs are, in their normal state, originally double."

Serres did not shrink from the decidedly peculiar logical implications of his solution. He noted that the large uterus had proper connections with the ovaries and vaginal canal and saw no reason why Ritta and Christina might not have borne children had they lived to maturity. (Serres also found a second, rudimentary uterus that would not have worked.) He concluded that the large uterus had formed half from Ritta and half from Christina, and admitted that any offspring developed within would have two natural mothers:

> This disposition of Ritta and Christina's genital organs evidently shows . . . that while nature had taken measures to assure the lives of these children, she had not forgotten the possibility of their reproduction. Now, for this reproduction, nature had combined everything, so that all the pleasures and pains would be shared. . . . Supposing that conception occurred in the large uterus, a single child would have had two distinct mothers, a singular result of this associated life.

Serres then discussed a pair of conjoined males with four legs and a single head, and opted for consistency and duality: the single well-formed brain shared the combined thoughts of two.

> There is a perfect unity produced by two distinct individualities. There are sense organs and cerebral hemispheres for a single individual, adapted to the service of two, since it is evident that there are two *me*'s in this single head [*deux* moi *dans cette tête unique*].

Thus Serres made a valiant and consistent attempt to resolve a question that seemed hopelessly ambiguous. We may appreciate the effort and enjoy an excursion into the different view of biology that Serres maintained. But we must reject his conclusion.

Fertilized human eggs usually develop into single individuals. Rarely, the dividing cells separate into discrete groups, and two embryos develop. These one-egg (or identical) twins are genetic carbon copies. In some ultimate, biological sense, they are the same iterated individual—and the psychological literature contains ample testimony to feelings of imperfect separation shared by many so-called identical twins. Yet, at least for definition's sake, we experience no difficulty in identifying one-egg human twins as undeniably separate personalities for two excellent reasons: first, physical separation is the essence of our vernacular definition of individuality (see following essay); second, human personalities are so subtly and pervasively shaped by complex environments of life (whatever the quirky similarities between one-egg twins reared apart) that each person follows a unique path.

With vastly greater rarity, the dividing cells of a fertilized egg begin to separate into two groups, but do not complete the process—and conjoined (or Siamese) twins develop. Conjoined twins span the entire conceivable range from a single individual bearing a few rudimentary parts of an imperfect twin, to superficially joined, complete individuals like Chang and Eng. Ritta and Christina fall squarely in the middle of this continuum. With our modern knowledge of their biological formation, I fear that we must reject Serres's solution, and admit instead that his dilemma cannot be answered.

We inhabit a complex world. Some boundaries are sharp and permit clean and definite distinctions. But nature also includes continua that cannot be neatly parceled into two piles of unambiguous yeses and noes. Biologists have rejected, as fatally flawed in principle, all attempts by anti-abortionists to define an unambiguous "beginning of life," because we know so well that the sequence from ovulation or spermatogenesis to birth is an unbreakable continuum—and surely no one will define masturbation as murder. Our congressmen may create a legal fiction for statutory effect but they may not seek support from biology. Ritta and Christina lay in the middle of another unbreakable continuum. They are in part two and in part one. And this, I am sorry to say, is the biological nonanswer to the question of the centuries.

If this argument leaves you with an empty feeling after so

much verbiage, I can only reply with the paradoxical phrase that is, so often, the most liberating response to an old mystery: The question has no answer because you asked the wrong question. The old question of individuality in Siamese twinning rests upon the assumption that objects can be pigeonholed into discrete categories. If we recognize that our world is full of irreducible continua, we will not be troubled by the intermediate status of Ritta and Christina.

Dante punished schismatics by dismembering them in hell to exact a physical punishment worthy of their ideological crime: "Lo, how is Mohammed mangled. . . . Whom here thou seest, while they lived, did sow scandal and schism, and therefore thus are rent."

Let us value connections. As Dante analogized physical with ideological separation, perhaps we can learn from the indissoluble union of Ritta and Christina that our intellectual world revels in continuity as well.

5 | A Most Ingenious Paradox

ABSTINENCE HAS its virtuous side, but enough is enough. I have always felt especially sorry for poor Mabel, betrothed to Frederic the pirate apprentice. On the very threshold of married happiness, she discovers that she must wait another sixty-three years to claim her beloved at age eighty-four—and, as could only happen in Gilbert and Sullivan, she actually promises to wait.

The Pirate King and Ruth, Frederic's old nursemaid and jilted paramour, present the reason for this extraordinary delay. Frederic, wrongfully apprenticed to the pirate band, has reached his twenty-first year and longs for freedom, respectability, and Mabel. But he was formally bound until his twenty-first *birthday,* and he was born on February 29. "You are a little boy of five," the Pirate King informs him with glee and expectation of prolonged service. The three principals of the *Pirates of Penzance* then analyze the complexities of this predicament in the famous paradox song:

> How quaint the ways of paradox
> At common sense she gaily mocks.

The classic paradox presents us with two contradictory interpretations, each quite correct in its own context. Consider our western prototypes, the so-called paradoxes of Zeno: The arrow that can never reach its destination because, at any instant, it must occupy a fixed position; and

Achilles who will never catch the tortoise because he must first traverse half the remaining distance, and any gap, no matter how small, can still be halved. We delight in paradox because it appeals to both the sublime and whimsical aspects of our psyche. We laugh with Frederic, but also feel that something deep about the nature of logic and life lies hidden in Zeno's conundrums.

Biology too has its classical paradox. It flared as a major issue in the nineteenth century, probably because scientists then felt that they might find a resolution. All the best naturalists struggled in vain: Huxley and Agassiz lined up on opposite sides; Haeckel tried to mediate. The twentieth century has largely bypassed the conundrum, probably because we now realize that no simple answer exists. Yet, if our fascination with paradox be justified, the question can still enlighten us by virtue of its stubborn intractability.

Physalia, the Portuguese man-of-war, embodies all this fuss. It is a siphonophore, a relative of corals and jellyfish. The old paradox addresses an issue that could not be more fundamental—the definition of an organism and the general question of boundaries in nature. Specifically: Are siphonophores organisms or colonies?

Siphonophores belong to the phylum Cnidaria (or Coelenterata). Two aspects of cnidarian biology set the context of our paradox. First, many cnidarians live as colonies of connected individuals—our massive coral reefs are gigantic congeries made of many million tiny, conjoined polyps. Second, the cnidarian life cycle features a so-called alternation of generations. The sessile polyp, a fixed cylinder with a fringe of tentacles, reproduces asexually and generates by budding the free-swimming medusa, or "jellyfish." The medusa produces sexual cells that unite and grow into a polyp. And so it goes.

Different kinds of cnidarians may emphasize one of these generations and suppress the other. Of the three major cnidarian groups, the Scyphozoa (or true jellyfish) have abandoned polyps and emphasized medusae, while the Anthozoa (or true corals) have dispensed with medusae and constructed reefs of polyps and their skeletons. In the third

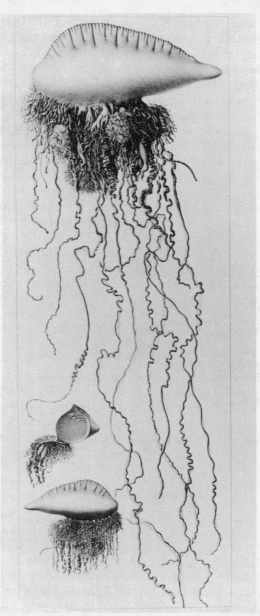

A Portuguese man-of-war. The creature is a colony, not a single organism. The float is a medusa person, and each "tentacle" is a polyp person. FROM LOUIS AGASSIZ'S MONOGRAPH (1862). REPRINTED FROM NATURAL HISTORY.

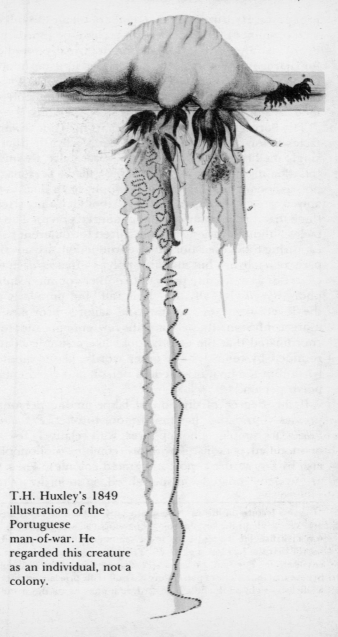

T.H. Huxley's 1849
illustration of the
Portuguese
man-of-war. He
regarded this creature
as an individual, not a
colony.

group, the Hydrozoa, many members retain the full cycle, with prominent polyp and medusa. Siphonophores are hydrozoans. The technical literature, not usually noted either for charm or directness, has transcended its usual limitatons in this case: amidst a forest of formidable jargon for other parts of cnidarian anatomy, it refers to the polyp and medusa stages of a single life cycle as "persons."

The Portuguese man-of-war, with its float above and tentacles below, looks superficially like a jellyfish (that is, a single medusa). When studied more carefully, we find that this floating weapon is a colony of many persons, both polyps and medusae. The pneumatophore, or float, is probably a greatly modified medusa (though some scientists think that it may be an even more altered polyp). The "tentacles," though varied and specialized for different roles of capturing food, digestion, and reproduction, are not simple parts of a jellyfish but modified polyps—that is, each tentacle arises as a discrete person. (Another common siphonophore, *Velella*, literally the "little sail" but popularly given the lovely name of "by-the-wind sailor," provokes even more confusion. Its persons are few enough and so well coordinated that the colony looks like a simple float surrounded by tentacles—in other words, like a simple jellyfish. But the float is a medusa person and each tentacle a polyp person.)*

If this degree of division of labor among persons impresses you, nature has much more to offer. *Physalia* and *Velella* are simple siphonophores, with relatively few types of modified persons. The more complex siphonophores are, by far, nature's most integrated colonies. Their parts are so differentiated and specialized, so subordinate to the

---

*I gave a lecture on this essay soon after its publication and drilled into my eager students the key phrase—"you thought a Portuguese man-of-war was a jellyfish, but it ain't." Later in the semester a student reported to my horror that she had lost a game of "Trivial Pursuit" by giving the *correct* answer to the question: "What is a Portuguese man-of-war?" Would you believe that the Solons of pop culture have officially proclaimed this colony a jellyfish—right on the little blue card, so it must be so. But it still ain't!

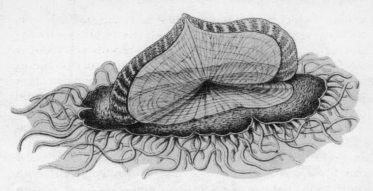

Velella, the "by-the-wind-sailor," is a colony—the float is a medusa person, each "tentacle" is a polyp person. FROM E. HAECKEL'S *Challenger* MONOGRAPH (1888). REPRINTED FROM NATURAL HISTORY.

entire colony, that they function more as organs of a body than persons of a colony.

Most siphonophores are small, transparent creatures of the open sea. They float at the surface among the plankton or swim actively, usually at shallow depths. As carnivores, they capture small planktonic animals in their net of tentacles. Larger siphonophores, *Physalia* among them, can ensnare and devour fish of substantial size; as many of us know to our sorrow, they can also inflict painful stings upon human bathers.

Complex siphonophores include an imposing array of well-differentiated structures. Their bodies may be roughly divided into two parts: an upper set of bulbs and pumps for locomotion and a lower set of tubes and filaments for feeding and reproduction. Each part contains a series of different polyps and medusae.

Consider first the range of forms and activities assumed by polyp persons. We find three basic types and a myriad of modifications. The feeding organs, or siphons (hence the group's name—*siphonophore* means "siphon bearer"), are tubular structures each with a stomach and trumpet-shaped mouth, usually hanging in profusion below the floats and

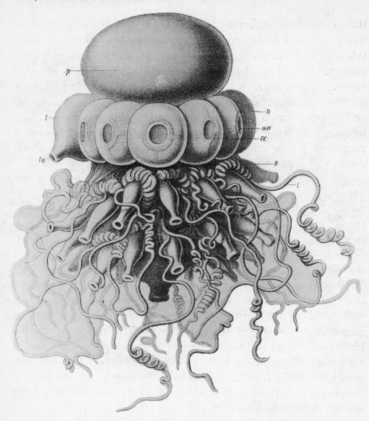

A relatively "simple" representative of the complex siphono-phores, just for starters. Only four basic types of individuals are shown—two upper kinds of medusa persons (the pneumatophore, or float, labeled p; and a row of nectophores, or swimming bells, labeled n); and two lower kinds of polyp persons (the feeding siphons, labeled s; and the long sensory tentacles, labeled t). FROM HAECKEL'S *Challenger* MONOGRAPH, 1888.

swimming persons. The siphons are minimally modified polyp persons, and we can easily comprehend their origin as complete organisms. All other types of polyps (and most medusae) are more highly altered and specialized, and therefore more difficult to link with their original personality. A second order of polyp persons, the so-called dactylozooids ("finger," or touching, animals), capture and transport food to the siphons. Dactylozooids include the extended thin tentacles, sometimes more than fifty feet long in *Physalia,* that carry the painful nematocysts, or stinging cells, and form a transparent web to ensnare prey. They have retained neither mouth nor digestive apparatus and might easily be taken for parts rather than persons if we could not trace their origin as discrete buds in growth.

These capturing parts often display a remarkable complexity of form and function. The stinging cells may be concentrated into swellings, or "batteries," sometimes protected by a hood. In *Stephanophyes,* each battery ends in a delicate terminal filament and contains about 1,700 stinging cells of four different types. The terminal filament lassoes the prey and discharges its few stinging cells. If these cells fail to dispatch the victim, the filament contracts and carries the prey to the far end of the battery itself, where another volley of larger stinging cells transfixes the victim. If the prey continues to struggle, another contraction moves it up the battery to the near end, where the largest and most powerful stinging cells finally end the torment before passing the vanquished prey along to the siphon for ingestion.

Jennifer E. Purcell (see bibliography) has recently presented further evidence that feeding and capturing persons do not form a simple passive network, like the web of a spider, but play an active role in obtaining food. She found that the stinging cell batteries of two species function as lures by resembling, in both form and motion, small zooplankton that serve as prey for animals eaten by siphonophores. The batteries of *Agalma okeni* look like a copepod with two long antennae; each contracts independently at varying intervals of five to thirty seconds, creating a web of motion that simulates the darting and swimming of a copepod school (or whatever you call an aggregation of these tiny planktonic arthropods). To seal the story, Purcell opened the stomachs of *Agalma* and found the remains of three creatures, all predators of copepods. The batteries of another species, *Athorybia rosacea,* resemble the planktonic larvae of fish. They also contract rapidly, mimicking the swimming and feeding motions of their models.

Gonozooids, the third category of polyp persons, are reproductive structures. They are usually short, simple tubes, without mouth or motion. But they bud off the medusa persons, which then make reproductive cells to produce the next generation of siphonophores.

The medusa persons of a complex siphonophore include four basic types: swimming, floating, protection, and reproduction. The swimming organs, or nectophores, are minimally modified medusae—basically the upper swimming bells without the lower tentacles. Some siphonophores carry several orderly rows of nectophores; their rhythmic muscular contractions propel the creature, often in elaborate, looping trajectories. The passive floats, or pneumatophores, are filled with gas (of a composition near ordinary air) and maintain the siphonophore at the surface or at some intermediate depth. Their origin is a matter of controversy. Long interpreted as modified medusa persons, some biologists now regard pneumatophores as even more elaborately transformed polyps. The two most familiar siphonophores, *Velella* and *Physalia,* build large floats but contain no nectophores. They therefore move passively on winds and currents, often drifting into bays and beaches in vast accumulations.

The covering organs, or bracts, are the most curiously modified structures of all. They are usually flat, shaped like a prism or a leaf, and so different in form and function from a medusa person that we would scarcely suspect their origin if we could not follow their growth and budding.

The reproductive medusae, or gonophores, are budded off from polyp persons, the gonozooids discussed earlier. In a few species, gonophores are liberated to float in the ocean as independent objects. But they cannot feed, and die soon after releasing their sex cells. In most siphonophores, however, gonophores never separate from the parent colony and remain attached as a kind of sexual organ.

The paradox of the Siphonophora expresses an issue that

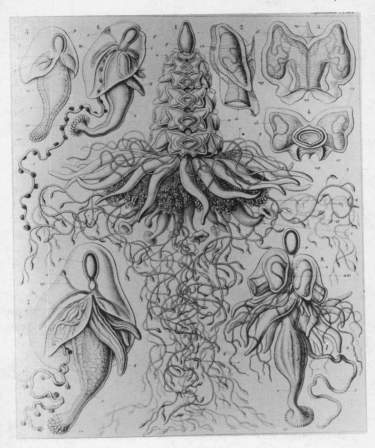

The middle figure shows a complete and complex siphonophore. The colony includes the following modified persons, from the top to the bottom: the single float, or pneumatophore (p); rows of swimming organs, or nectophores (n); fingerlike sensory projections, or palpons (q); clusters of reproductive parts (g); feeding siphons with trumpet-shaped mouths (s); and long, twisted strands of food-capturing filaments (t). Other figures are parts or early growth stages of the complex colony. FROM E. HAECKEL'S *Challenger* MONOGRAPH, 1888. REPRINTED FROM NATURAL HISTORY.

One more complex siphonophore for good measure, and with yet another kind of person added (protective bracts). From top to bottom: a single pneumatophore; two vertical rows of necto-phores; protective leaflike bracts; feeding siphons with trumpet-shaped mouths; and finally, long food-capturing filaments. FROM E. HAECKEL (1888). REPRINTED FROM NATURAL HISTORY.

I have been avoiding, or rather skirting about, in presenting this taxonomy of persons or parts. I have described the various swimming, floating, protecting, feeding, capturing, and reproducing structures as persons—that is, as individual polyp or medusa organisms. Using evolutionary history as a criterion, this designation is almost surely correct and accepted by nearly all biologists. By history, siphonophores are colonies; they evolved from simpler aggregations of discrete organisms, each reasonably complete and able to perform a nearly full set of functions (as in modern coral colonies). But the colony has become so integrated, and the different persons so specialized in form and so subordinate to the whole, that the entire aggregation now functions as a single individual, or superorganism.

The persons of a siphonophore no longer maintain individuality in a functional sense. They are specialized for a single task and perform as organs of a larger entity. They do not look like organisms and could not survive as separate creatures. The entire colony works as a single being, and its parts (or persons) move in a coordinated manner. Although each nectophore (or swimming bell) maintains its own nervous system, a common nerve tract connects the entire set. Impulses along this pathway regulate the rows of nectophores in an integrated manner that permits the whole colony (or animal) to move with precision and grace. Touch the float of *Nanomia* at one end, and nectophores at the other extremity contract to remove the animal (or colony, if you will) from danger. Siphons pump their digested food along the common stem to the rest of the colony, but empty siphons also join in the general peristalsis, and food, as a result, reaches the entire colony (or organism) more effectively.

My studied parentheticals of the last paragraph underscore the fundamental paradox. Shall we call the entire siphonophore a colony or an organism—for it is a colony by evolutionary history but more an organism by current function. And what of the parts or persons? By history, they are modified individuals; by current function, they are organs of a larger entity. What is to be done?

This issue fueled the great siphonophore debate of nineteenth-century natural history. T.H. Huxley studied siphonophores during his long apprenticeship at sea on H.M.S. *Rattlesnake* (less celebrated than Darwin's adventure on the *Beagle,* but another example of the same extended, exemplary, and largely extinct style of training in natural history). He interpreted siphonophores as conventional organisms, their parts as true organs and not modified persons. Huxley used siphonophores as his primary example in a famous essay on the nature of individuality in biology.

Louis Agassiz studied the Portuguese man-of-war on the shores of his adopted America. (I have included his beautiful lithograph of *Physalia* with this essay) and decided that siphonophores are colonies, their integration a sign of divine handiwork.

Ernst Haeckel, artist and naturalist *extraordinaire*, described the siphonophores collected on that most celebrated of scientific expeditions in oceanography, the voyage of H.M.S. *Challenger*, 1873–1876. He published with his report a series of plates (including all other illustrations in this essay), unmatched ever since for beauty (though a bit short on accuracy, since Haeckel often added a touch of heightened symmetry for artistic effect). Haeckel also included several plates of siphonophores in his *Kunstformen der Natur (Art Forms in Nature)* of 1904—the great series of 100 lithographs, with plants and animals arranged in weirdly distorted form and swirling symmetry, in the best tradition of reigning *art nouveau* so well embodied in contemporary kiosks of the Paris Métro.

Haeckel's theory of siphonophores would require an essay in itself to explain and explore, but he tried to mediate between Huxley and Agassiz by viewing these creatures in part as colonies (the poly-person theory in his words), in part as organisms (the poly-organ theory). Haeckel also used siphonophores, as Huxley had, to illustrate by dubious analogy his views on the proper organization of human societies. In his *Über Arbeitstheilung in Natur und Menschenleben (On the Division of Labor in Nature and Human Life)*, he compared the simple colonies of other cnidarians with the life styles of "primitive" humans and their limited division of labor for repetitive tasks performed by all: "The wild people of nature, who have remained on the lowest level right to our own day, lack both culture and division of labor—or they limit division of labor, as do most animals, to the different tasks of the two sexes." He then compared complex colonies of siphonophores with the "advances" that division of labor permits in "higher" human societies—including modern warfare, where instruments of destruction "require hundreds of human hands, working in different ways and manners."

Can we now suggest any resolution to this ancient debate, any possible mediation between two legitimate criteria that

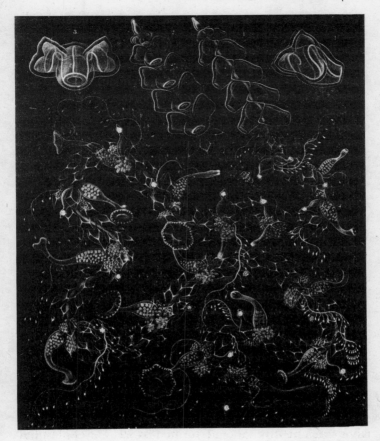

And finally, just for its aesthetic value, another Haeckel (1888) plate of complex siphonophores.

seem to yield opposite results—the criterion of history supporting the poly-person theory (siphonophores are colonies and their parts are persons) and the criterion of current function upholding the poly-organ theory (siphonophores are organisms and their parts are organs)? Can we tip the balance in favor of one view or the other by invoking the third major criterion of natural history—growth and form?

Growth and form provide us with an *embarras de richesses* by presenting evidence for and against both theories. As strong support for the poly-organ theory, siphonophores develop from a single fertilized egg cell. A siphonophore begins life as an unambiguous person—should we not regard any later development as an elaboration of this founding individual? Moreover, the adult siphonophore acts as a discrete object. Many species exhibit definite and complex symmetry governing all parts considered together. Some Portuguese men-of-war, for example, come in right- and left-handed versions.

We may, however, cite equally good arguments for the poly-person theory. Admittedly, each colony begins life as a single ovum, but it then develops a series of entities—full persons in this view—by budding from a common stem. This mode of growth is familiar in many aggregations conventionally regarded as colonies. A stand of bamboo may trace its origin to a single seed, yet we usually view each budded stem as an individual.

In addition, highly specialized structures sometimes bear vestigial parts that testify to their status as persons. In the poly-person theory, for example, nectophores are medusae that have lost all feeding and digestive parts, retaining only the jellyfish bell. But some nectophores grow rudimentary tentacles; in one species, the tentacles even retain eyespots. Protective bracts are the most modified and specialized of all siphonophore parts, but the bracts of two species retain a vestigial mouth—an indication that they arose as full medusa persons.

It looks like a tossup again. We might resolve our paradox if growth occurred in either of two ways—but nature doesn't oblige. If all structures began growth as complete persons with a full set of parts, and then lost unneeded pieces as they specialized for swimming, protecting, or eating, then the poly-person theory would gain a big boost. If buds from the main stem began as complete persons and then disarticulated—the bell parts becoming nectophores and the tentacle parts siphons, for example—then the poly-organ theory would be affirmed. But most specialized parts simply grow as we find them. Nectophores differentiate as nectophores, bracts as bracts. We are immersed in an unresolvable conflict among equally legitimate criteria: discrete buds grow like a person with specialized parts like an organ. What, for example, shall we make of a gonophore, the degenerate reproductive medusa budded from a polyp? If it separates from the colony, we may choose to regard the gonophore as an organism. But it has no mouth and cannot feed; it must therefore die after releasing the sexual cells. Should we call such a limited breeding machine an individual? And if the gonophore remains attached to the colony, as it usually does, should we regard it as any more than a sexual organ?

When an inquiry becomes so convoluted, we must suspect that we are proceeding in the wrong way. We must return to go, change gears, and reformulate the problem, not pursue every new iota of information or nuance of argument in the old style, hoping all the time that our elusive solution simply awaits a crucial item, yet undiscovered.

Nature, in some respects, comes to us as continua, not as discrete objects with clear boundaries. One of nature's many continua extends from colonies at one end to organisms at the other. Even the basic terms—organism and colony—have no precise and unambiguous definitions. We may, however, use the two criteria of our vernacular as a guide. We tend to call a biological object an organism if it maintains no permanent physical connection with others and if its parts are so well integrated that they work only in coordination and for the proper function of the whole.

Most creatures lie near one or the other end of this continuum, and we have no trouble defining them as organisms or colonies. People are organisms—even though all multicellular creatures probably arose as colonies about a billion years ago. This origin is so distant, and so much has happened since, that we detect no signs of coloniality in our current operation. Thus, we are organisms by any reasonable use of language. Reef-building corals are colonies because each polyp is a complete creature, fully functional in its own right, though attached to its fellows.

But since nature has built a continuum from colony to organism, we must encounter ambiguity at the center. Some cases will be impossible to call—as a property of nature, not an imperfection of knowledge. Consider a progression from evident organisms toward the undefinable center. Human societies are made of organisms; each person is genetically distinct and spatially separate. What about ants? We still opt for organisms even though ants may so submerge their individuality in tightly knit societies that some naturalists refer to an ant colony as a superorganism.

What about aphids? We begin to lose clarity. All members of an aphid clone are female; each founding mother grows her young within her own body, without benefit of fertilization. All her offspring are genetically identical. Is the clone an aggregation of separate individuals or one gigantic evolutionary body with many thousand separate parts, all identical? (One prominent evolutionary biologist has recently urged this second view.)

What about a stand of bamboo? Harder still. All stems are members of a clone; they are genetically identical, and attached to a common underground runner. Is each plant above ground a person or a part? We still usually opt for persons (though many biologists demur) because each plant looks much the same and has a full set of structures.*

Finally, then, what about siphonophores? We are now squarely in the middle of a continuum, and we cannot provide a clear answer. The parts of siphonophores are persons by history, organs by current function, and a bit of both by growth. Our criteria of separation and independent operation have failed, but we cannot deny a history that still stares us in the face.

---

*Since botanists face this dilemma more often than zoologists, they have devised a terminology for these ambiguous cases—"genet" for the entire aggregation and "ramet" for each iterated set of parts. This new terminology is no solution, but simply a formal recognition that the issue cannot be resolved with our usual concepts of individuality.

Siphonophores do not convey the message—a favorite theme of unthinking romanticism—that nature is but one gigantic whole, all its parts intimately connected and interacting in some higher, ineffable harmony. Nature revels in boundaries and distinctions; we inhabit a universe of structure. But since our universe of structure has evolved historically, it must present us with fuzzy boundaries, where one kind of thing grades into another. Objects at these boundaries will continue to confuse and frustrate us so long as we follow old habits of thought and insist that all parts of nature be pigeonholed unambiguously to assuage our poor and overburdened intellects.

The siphonophore paradox does have an answer of sorts, and a profound one at that. The answer is that we asked the wrong question—a question that has no meaning because its assumptions violate the ways of nature. Are siphonophores organisms or colonies? Both and neither; they lie in the middle of a continuum where one grades into the other.

The siphonophore paradox is illuminating, not discouraging. It cannot be resolved, but when we understand why, we grasp a great truth about nature's structure. Siphonophores deliver the same message as that old one about the lady who visits her butcher one Friday morning, seeking a large chicken for the Sabbath meal. The butcher looks in his bin and finds to his chagrin that he has but one scrawny animal left. He takes it out with great fanfare and puts it on the scale. Two pounds. "Not big enough," the lady says. He puts it back in the bin, pretends to rummage amidst a large pile of nonexistent alternatives, finally pulls out the same chicken, puts it on the scale, and puts his thumb on the scale. Three pounds. "Fine," says the lady. "I'll take them both."* Things that seem separate are often the different sides of a unity.

*Dr. S.I. Joseph has since told me that he saw the same lady at a fruit stand later that day. She was asking about the price of grapefruit. "Two for thirty-five cents" she learned. "How much for one," she asked. "Twenty cents" came the reply. "Fine," she said, "I'll take the other."

2 | Theory and Perception

6 | Adam's Navel

THE AMPLE FIG LEAF served our artistic forefathers well as a botanical shield against indecent exposure for Adam and Eve, our naked parents in the primeval bliss and innocence of Eden. Yet, in many ancient paintings, foliage hides more than Adam's genitalia; a wandering vine covers his navel as well. If modesty enjoined the genital shroud, a very different motive—mystery—placed a plant over his belly. In a theological debate more portentous than the old argument about angels on pinpoints, many earnest people of faith had wondered whether Adam had a navel.

He was, after all, not born of a woman and required no remnant of his nonexistent umbilical cord. Yet, in creating a prototype, would not God make his first man like all the rest to follow? Would God, in other words, not create with the appearance of preexistence? In the absence of definite guidance to resolve this vexatious issue, and not wishing to incur anyone's wrath, many painters literally hedged and covered Adam's belly.

A few centuries later, as the nascent science of geology gathered evidence for the earth's enormous antiquity, some advocates of biblical literalism revived this old argument for our entire planet. The strata and their entombed fossils surely seem to represent a sequential record of countless years, but wouldn't God create his earth with the appearance of preexistence? Why should we not believe that he created strata and fossils to give modern life a harmonious

99

order by granting it a sensible (if illusory) past? As God provided Adam with a navel to stress continuity with future men, so too did he endow a pristine world with the appearance of an ordered history. Thus, the earth might be but a few thousand years old, as Genesis literally affirmed, and still record an apparent tale of untold eons.

This argument, so often cited as a premier example of reason at its most perfectly and preciously ridiculous, was most seriously and comprehensively set forth by the British naturalist Philip Henry Gosse in 1857. Gosse paid proper homage to historical context in choosing a title for his volume. He named it *Omphalos* (Greek for navel), in Adam's honor, and added as a subtitle: *An Attempt to Untie the Geological Knot.*

Since *Omphalos* is such spectacular nonsense, readers may rightly ask why I choose to discuss it at all. I do so, first of all, because its author was such a serious and fascinating man, not a hopeless crank or malcontent. Any honest passion merits our attention, if only for the oldest of stated reasons—Terence's celebrated *Homo sum: humani nihil a me alienum puto* (I am human, and am therefore indifferent to nothing done by humans).

Philip Henry Gosse (1810–1888) was the David Attenborough of his day, Britain's finest popular narrator of nature's fascination. He wrote a dozen books on plants and animals, lectured widely to popular audiences, and published several technical papers on marine invertebrates. He was also, in an age given to strong religious feeling as a mode for expressing human passions denied vent elsewhere, an extreme and committed fundamentalist of the Plymouth Brethren sect. Although his *History of the British Sea-Anemones* and other assorted ramblings in natural history are no longer read, Gosse retains some notoriety as the elder figure in that classical work of late Victorian self-analysis and personal exposé, his son Edmund's wonderful account of a young boy's struggle against a crushing religious extremism imposed by a caring and beloved parent— *Father and Son.*

My second reason for considering *Omphalos* invokes the

same theme surrounding so many of these essays about nature's small oddities: Exceptions do prove rules (prove, that is, in the sense of probe or test, not affirm). If you want to understand what ordinary folks do, one thoughtful deviant will teach you more than ten thousand solid citizens. When we grasp why *Omphalos* is so unacceptable (and not, by the way, for the reason usually cited), we will understand better how science and useful logic proceed. In any case, as an exercise in the anthropology of knowledge, *Omphalos* has no parallel—for its surpassing strangeness arose in the mind of a stolid Englishman, whose general character and cultural setting we can grasp as akin to our own, while the exotic systems of alien cultures are terra incognita both for their content and their context.

To understand *Omphalos,* we must begin with a paradox. The argument that strata and fossils were created all at once with the earth, and only present an illusion of elapsed time, might be easier to appreciate if its author had been an urban armchair theologian with no feeling or affection for nature's works. But how could a keen naturalist, who had spent days, nay months, on geological excursions, and who had studied fossils hour after hour, learning their distinctions and memorizing their names, possibly be content with the prospect that these objects of his devoted attention had never existed—were, indeed, a kind of grand joke perpetrated upon us by the Lord of All?

Philip Henry Gosse was the finest descriptive naturalist of his day. His son wrote: "As a collector of facts and marshaller of observations, he had not a rival in that age." The problem lies with the usual caricature of *Omphalos* as an argument that God, in fashioning the earth, had consciously and elaborately lied either to test our faith or simply to indulge in some inscrutable fit of arcane humor. Gosse, so fiercely committed both to his fossils and his God, advanced an opposing interpretation that commanded us to study geology with diligence and to respect all its facts even though they had no existence in real time. When we understand why a dedicated empiricist could embrace the argument of *Omphalos* ("creation with the appearance of preexistence"), only then can we understand its deeper fallacies.

Gosse began his argument with a central, but dubious, premise: All natural processes, he declared, move endlessly round in a circle: egg to chicken to egg, oak to acorn to oak.

> This, then, is the order of all organic nature. When once we are in any portion of the course, we find ourselves running in a circular groove, as endless as the course of a blind horse in a mill. . . . [In premechanized mills, horses wore blinders or, sad to say, were actually blinded, so that they would continue to walk a circular course and not attempt to move straight forward, as horses relying on visual cues tend to do.] This is not the law of some particular species, but of all: it pervades all classes of animals, all classes of plants, from the queenly palm down to the protococcus, from the monad up to man: the life of every organic being is whirling in a ceaseless circle, to which one knows not how to assign any commencement. . . . The cow is as inevitable a sequence of the embryo, as the embryo is of the cow.

When God creates, and Gosse entertained not the slightest doubt that all species arose by divine fiat with no subsequent evolution, he must break (or "erupt," as Gosse wrote) somewhere into this ideal circle. Wherever God enters the circle (or "places his wafer of creation," as Gosse stated in metaphor), his initial product must bear traces of previous stages in the circle, even if these stages had no existence in real time. If God chooses to create humans as adults, their hair and nails (not to mention their navels) testify to previous growth that never occurred. Even if he decides to create us as a simple fertilized ovum, this initial form implies a phantom mother's womb and two nonexistent parents to pass along the fruit of inheritance.

> Creation can be nothing else than a series of irruptions into circles. . . . Supposing the irruption to have been made at what part of the circle we please, and

varying this condition indefinitely at will,—we cannot avoid the conclusion that each organism was from the first marked with the records of a previous being. But since creation and previous history are inconsistent with each other; as the very idea of the creation of an organism excludes the idea of pre-existence of that organism, or of any part of it; it follows, that such records are false, so far as they testify to time.

Gosse then invented a terminology to contrast the two parts of a circle before and after an act of creation. He labeled as "prochronic," or occurring outside of time, those appearances of preexistence actually fashioned by God at the moment of creation but seeming to mark earlier stages in the circle of life. Subsequent events occurring after creation, and unfolding in conventional time, he called "diachronic." Adam's navel was prochronic, the 930 years of his earthly life diachronic.

Gosse devoted more than 300 pages, some 90 percent of his text, to a simple list of examples for the following small part of his complete argument—if species arise by sudden creation at any point in their life cycle, their initial form must present illusory (prochronic) appearances of preexistence. Let me choose just one among his numerous illustrations, both to characterize his style of argument and to present his gloriously purple prose. If God created vertebrates as adults, Gosse claimed, their teeth imply a prochronic past in patterns of wear and replacement.

Gosse leads us on an imaginary tour of life just an hour after its creation in the wilderness. He pauses at the seashore and scans the distant waves:

I see yonder a . . . terrific tyrant of the sea. . . . It is the grisly shark. How stealthily he glides along. . . . Let us go and look into his mouth. . . . Is not this an awful array of knives and lancets? Is not this a case of surgical instruments enough to make you shudder? What would be the amputation of your leg to this row of triangular scalpels?

Yet the teeth grow in spirals, one behind the next, each waiting to take its turn as those in current use wear down and drop out:

> It follows, therefore, that the teeth which we now see erect and threatening, are the successors of former ones that have passed away, and that they were once dormant like those we see behind them. . . . Hence we are compelled by the phenomena to infer a long past existence to this animal, which yet has been called into being within an hour.

Should we try to argue that teeth in current use are the first members of their spiral, implying no predecessors after all, Gosse replies that their state of wear indicates a pro-chronic past. Should we propose that these initial teeth might be unmarred in a newly created shark, Gosse moves on to another example.

> Away to a broader river. Here wallows and riots the huge hippopotamus. What can we make of his dentition?

All modern adult hippos possess strongly worn and beveled canines and incisors, a clear sign of active use throughout a long life. May we not, however, as for our shark, argue that a newly created hippo might have sharp and pristine front teeth? Gosse argues correctly that no hippo could work properly with teeth in such a state. A created adult hippo must contain worn teeth as witnesses of a prochronic past:

> The polished surfaces of the teeth, worn away by mutual action, afford striking evidence of the lapse of time. Some one may possibly object. . . . "What right have you to assume that these teeth were worn away at the moment of its creation, admitting the animal to have been created adult. May they not have been entire?" I reply, Impossible: the Hippopotamus's teeth

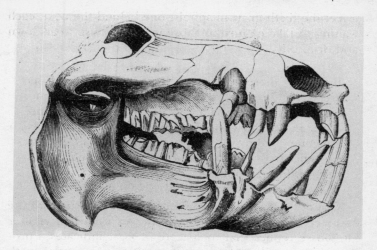

Gosse's (1857) figure of a hippo's skull showing the beveled tooth edges (a product of time and wear) necessary for function.

would have been perfectly useless to him, except in the ground-down condition: nay, the unworn canines would have effectually prevented his jaws from closing, necessitating the keeping of the mouth wide open until the attrition was performed; long before which, of course, he would have starved. . . . The degree of attrition is merely a question of time. . . . How distinct an evidence of past action, and yet, in the case of the created individual, how illusory!

This could go on forever (it nearly does in the book), but just one more dental example. Gosse, continuing upward on the topographic trajectory of his imaginary journey, reaches an inland wood and meets *Babirussa,* the famous Asian pig with upper canines growing out and arching back, almost piercing the skull:

In the thickets of this nutmeg grove beside us there is a Babiroussa; let us examine him. Here he is, almost

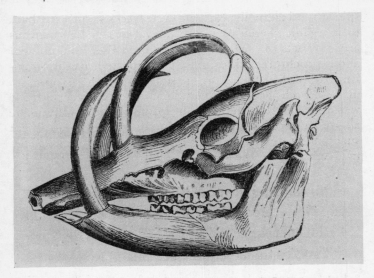

A *Babirussa* skull from Gosse (1857) showing implied time in wear of the molars and growth of the arched canines.

submerged in this tepid pool. Gentle swine with the circular tusk, please to open your pretty mouth!

The pig, created by God but an hour ago, obliges, thus displaying his worn molars and, particularly, the arching canines themselves, a product of long and continuous growth.

I find this part of Gosse's argument quite satisfactory as a solution, within the boundaries of his assumptions, to that classical dilemma of reasoning (comparable in importance to angels on pinpoints and Adam's navel): "Which came first, the chicken or the egg?" Gosse's answer: "Either, at God's pleasure, with prochronic traces of the other." But arguments are only as good as their premises, and Gosse's inspired nonsense fails because an alternative assumption, now accepted as undoubtedly correct, renders the question irrelevant—namely, evolution itself. Gosse's circles do not spin around eternally; each life cycle traces an ancestry back to inorganic chemicals in a primeval ocean. If organisms arose by acts of creation *ab nihilo,* then Gosse's argument about prochronic traces must be respected. But if organisms evolved to their current state, *Omphalos* collapses to massive irrelevance. Gosse understood this threat perfectly well and chose to meet it by abrupt dismissal. Evolution, he allowed, discredited his system, but only a fool could accept such patent nonsense and idolatry (Gosse wrote *Omphalos* two years before Darwin published the *Origin of Species*).

> If any choose to maintain, as many do, that species were gradually brought to their present maturity from humbler forms . . . he is welcome to his hypothesis, but I have nothing to do with it. These pages will not touch him.

But Gosse then faced a second and larger difficulty: The prochronic argument may work for organisms and their life cycles, but how can it be applied to the entire earth and its fossil record—for Gosse intended *Omphalos* as a treatise to reconcile the earth with biblical chronology, "an attempt to untie the geological knot." His statements about prochronic parts in organisms are only meant as collateral support for the primary geological argument. And Gosse's geological claim fails precisely because it rests upon such dubious analogy with what he recognized (since he gave it so much more space) as a much stronger argument about modern organisms.

Gosse tried valiantly to advance for the entire earth the same two premises that made his prochronic argument work for organisms. But an unwilling world rebelled against such forced reasoning and *Omphalos* collapsed under its own weight of illogic. Gosse first tried to argue that all geological processes, like organic life cycles, move in circles:

> The problem, then, to be solved before we can certainly determine the question of analogy between the globe and the organism, is this—Is the life-history of

the globe a cycle? If it is (and there are many reasons why this is probable), then I am sure prochronism must have been evident at its creation, since there is no point in a circle which does not imply previous points.

But Gosse could never document any inevitable geological cyclicity, and his argument drowned in a sea of rhetoric and biblical allusion from Ecclesiastes: "All the rivers run into the sea; yet the sea *is* not full. Unto the place from whence the rivers come, thither they return again."

Secondly, to make fossils prochronic, Gosse had to establish an analogy so riddled with holes that it would make the most ardent mental tester shudder—embryo is to adult as fossil is to modern organism. One might admit that chickens require previous eggs, but why should a modern reptile (especially for an antievolutionist like Gosse) be necessarily linked to a previous dinosaur as part of a cosmic cycle? A python surely does not imply the ineluctable entombment of an illusory *Triceratops* into prochronic strata.

With this epitome of Gosse's argument, we can resolve the paradox posed at the outset. Gosse could accept strata and fossils as illusory and still advocate their study because he did not regard the prochronic part of a cycle as any less "true" or informative than its conventional diachronic segment. God decreed two kinds of existence—one constructed all at once with the appearance of elapsed time, the other progressing sequentially. Both dovetail harmoniously to form uninterrupted circles that, in their order and majesty, give us insight into God's thoughts and plans.

The prochronic part is neither a joke nor a test of faith; it represents God's obedience to his own logic, given his decision to order creation in circles. As thoughts in God's mind, solidified in stone by creation *ab nihilo,* strata and fossils are just as true as if they recorded the products of conventional time. A geologist should study them with as much care and zeal, for we learn God's ways from both his prochronic and his diachronic objects. The geological time scale is no more meaningful as a yardstick than as a map of God's thoughts.

> The acceptance of the principles presented in this volume . . . would not, in the least degree, affect the study of scientific geology. The character and order of the strata; . . . the successive floras and faunas; and all the other phenomena, would be facts still. They would still be, as now, legitimate subjects of examination and inquiry. . . . We might still speak of the inconceivably long duration of the processes in question, provided we understand ideal instead of actual time—that the duration was projected in the mind of God, and not really existent.

Thus, Gosse offered *Omphalos* to practicing scientists as a helpful resolution of potential religious conflicts, not a challenge to their procedures or the relevance of their information.

His son Edmund wrote of the great hopes that Gosse held for *Omphalos*:

> Never was a book cast upon the waters with greater anticipations of success than was this curious, this obstinate, this fanatical volume. My father lived in a fever of suspense, waiting for the tremendous issue. This *"Omphalos"* of his, he thought, was to bring all the turmoil of scientific speculation to a close, fling geology into the arms of Scripture, and make the lion eat grass with the lamb.

Yet readers greeted *Omphalos* with disbelief, ridicule, or worse, stunned silence. Edmund Gosse continued:

> He offered it, with a glowing gesture, to atheists and Christians alike. This was to be the universal panacea; this the system of intellectual therapeutics which could not but heal all the maladies of the age. But, alas!

atheists and Christians alike looked at it and laughed, and threw it away.

Although Gosse reconciled himself to a God who would create such a minutely detailed, illusory past, this notion was anathema to most of his countrymen. The British are a practical, empirical people, "a nation of shopkeepers" in Adam Smith's famous phrase; they tend to respect the facts of nature at face value and rarely favor the complex systems of nonobvious interpretation so popular in much of continental thought. Prochronism was simply too much to swallow. The Reverend Charles Kingsley, an intellectual leader of unquestionable devotion to both God and science, spoke for a consensus in stating that he could not "give up the painful and slow conclusion of five and twenty years' study of geology, and believe that God has written on the rocks one enormous and superfluous lie."

And so it has gone for the argument of *Omphalos* ever since. Gosse did not invent it, and a few creationists ever since have revived it from time to time. But it has never been welcome or popular because it violates our intuitive notion of divine benevolence as free of devious behavior— for while Gosse saw divine brilliance in the idea of prochronism, most people cannot shuck their seat-of-the-pants feeling that it smacks of plain old unfairness. Our modern American creationists reject it vehemently as imputing a dubious moral character to God and opt instead for the even more ridiculous notion that our miles of fossiliferous strata are all products of Noah's flood and can therefore be telescoped into the literal time scale of Genesis.

But what is so desperately wrong with *Omphalos?* Only this really (and perhaps paradoxically): that we can devise no way to find out whether it is wrong—or, for that matter, right. *Omphalos* is the classical example of an utterly untestable notion, for the world will look exactly the same in all its intricate detail whether fossils and strata are prochronic or products of an extended history. When we realize that *Omphalos* must be rejected for this methodological absurdity, not for any demonstrated factual inaccuracy, then we will understand science as a way of knowing, and *Omphalos* will serve its purpose as an intellectual foil or prod.

Science is a procedure for testing and rejecting hypotheses, not a compendium of certain knowledge. Claims that can be proved incorrect lie within its domain (as false statements to be sure, but as proposals that meet the primary methodological criterion of testability). But theories that cannot be tested in principle are not part of science. Science is doing, not clever cogitation; we reject *Omphalos* as useless, not wrong.

Gosse's deep error lay in his failure to appreciate this essential character of scientific reasoning. He hammered his own coffin nails by continually emphasizing that *Omphalos* made no practical difference—that the world would look exactly the same with a prochronic or diachronic past. (Gosse thought that this admission would make his argument acceptable to conventional geologists; he never realized that it could only lead them to reject his entire scheme as irrelevant.) "I do not know," he wrote, "that a single conclusion, now accepted, would need to be given up, except that of actual chronology."

Gosse emphasized that we cannot know where God placed his wafer of creation into the cosmic circle because prochronic objects, created *ab nihilo,* look exactly like diachronic products of actual time. To those who argued that coprolites (fossil excrement) prove the existence of active, feeding animals in a real geological past, Gosse replied that as God would create adults with feces in their intestines, so too would he place petrified turds into his created strata. (I am not making up this example for comic effect; you will find it on page 353 of *Omphalos* .) Thus, with these words, Gosse sealed his fate and placed himself outside the pale of science:

> Now, again I repeat, there is no imaginable difference to sense between the prochronic and the diachronic development. Every argument by which the physiologist can prove to demonstration that yonder cow was once a foetus in the uterus of its dam, will apply with exactly the same power to show that the newly created

cow was an embryo, some years before its creation. . . . There is, and can be, nothing in the phenomena to indicate a commencement there, any more than anywhere else, or indeed, anywhere at all. The commencement, as a fact, I must learn from testimony; I have no means whatever of inferring it from phenomena.

Gosse was emotionally crushed by the failure of *Omphalos*. During the long winter evenings of his discontent, in the January cold of 1858, he sat by the fire with his eight-year-old son, trying to ward off bitter thoughts by discussing the grisly details of past and current murders. Young Edmund heard of Mrs. Manning, who buried her victim in quicklime and was hanged in black satin; of Burke and Hare, the Scottish ghouls; and of the "carpetbag mystery," a sackful of neatly butchered human parts hung from a pier on Waterloo Bridge. This may not have been the most appropriate subject for an impressionable lad (Edmund was, by his own memory, "nearly frozen into stone with horror"), yet I take some comfort in the thought that Philip Henry Gosse, smitten with the pain of rejection for his untestable theory, could take refuge in something so unambiguously factual, so utterly concrete.

Postscript

I have since learned that one of my favorite writers, Jorge Luis Borges, wrote a fascinating short comment on *Omphalos* ("The creation and P.H. Gosse" in *Other Inquisitions, 1937–1952*, published in 1964 by the University of Texas Press, Ruth L.C. Simms, translator). Borges begins by citing several literary references to the absence of a navel in our primal parents. Sir Thomas Browne, as a metaphor for original sin, writes in *Religio Medici* (1642), "the man without a navel yet lives in me"; and James Joyce, in the first chapter

of *Ulysses* (what cannot be found in this incredible book!)
states, "Heva, naked Eve. She had no navel." I particularly
appreciated the lovely epitome and insight of Borges's conclusion (though I disagree with his second point), "I should
like to emphasize two virtues in Gosse's forgotten thesis.
First, its rather monstrous elegance. Second: its involuntary
reduction of a *creatio ab nihilo* to absurdity, its indirect demonstration that the universe is eternal, as the Vedanta,
Heraclitus, Spinoza and the atomists thought."

7 | The Freezing of Noah

TIDBITS OF A DISTANT PAST often reemerge into our present with surprising relevance. After all, human thought and emotion have a universality that transcends time and converts the different stages of history into theaters that provide lessons for modern players.

I want to tell a tale of twenty years in the history of British geology—roughly 1820 to 1840. The story displays science working at its best. One of Britain's premier geologists proposed a theory. This clearly formulated statement had roots (as do all theories) in the social position and psychological constitution of its founder. But it was also empirically based and eminently testable. The theory was tested, and it failed. Its two primary supporters recanted forthrightly and later led an effort to formulate different and more adequate explanations for the phenomena that had inspired the original theory.

In 1823, the Reverend William Buckland (1784–1856), Oxford University's first "official" geologist, published a scientific treatise with a striking title that reflected its author's attempt to amalgamate his two professional worlds—religion and geology. He called it *Reliquiae diluvianae,* or *Relics of the Flood.* Its subtitle indicated the kind of evidence that Buckland would cite to support his theory about the geological expression of Noah's debacle: *Observations on the Organic Remains Contained in Caves, Fissures, and Diluvial Gravel, and on Other Geological Phenomena Attesting the Action of a Universal Deluge.* Buckland's theory was tested and rejected by geologists who were both creationists and genuine scientists. Noah's flood has not been an issue among geologists for the past century and a half.

Those modern fundamentalists who call themselves "scientific creationists" have resurrected Noah and made his flood the linchpin of their system. In fact, they ascribe *all* fossil-bearing strata to this single event, whereas Buckland, much more sensibly, sought only to identify the uppermost, unconsolidated film of loams and gravels as products of Noah's universal deluge. The recognition of Noah's flood as a primary geological agent was specifically mandated in the Arkansas "creation science" law, declared unconstitutional in January 1982. I know no better illustration of the difference between science and pseudoscience than a comparison of Buckland's rational approach—concrete proposal, test, and rejection—with the dogmatism of fundamentalists.

Buckland was not the first geologist to propose a "flood theory" linking Noah's deluge to the evidence of geology, but his new version had the twin virtues of sensibility and testability. The granddaddy of flood theories (and the one now embraced so anachronistically by creationists) had been kicking around for several centuries—the idea that a single flood had produced all, or nearly all, the geological strata. This version was no longer credible by Buckland's time, and he dismissed it in a single paragraph written in 1836 and still quite sufficient to refute what our moral majoritarians tried to impose upon the children of Arkansas:

> Some have attempted to ascribe the formation of all the stratified rocks to the effects of the Mosaic Deluge; an opinion which is irreconcilable with the enormous thickness and almost infinite subdivisions of these strata, and with the numerous and regular successions which they contain of the remains of animals and vegetables, differing more and more widely from existing species, as the strata in which we find them are older, or placed at greater depths.

Other geologists had viewed the Flood as a time of upheaval on the earth's surface. Old lands foundered, while new continents rose from the oceanic depths—thus explaining the presence of fossil shells on mountaintops. But Buckland recognized that the earth had an ancient history, punctuated sporadically (but often) with episodes of uplift. He needed no recent flood to explain the earth's topography and the geological contents of its mountains.

Buckland's flood was a less eventful, less catastrophic, and much more believable episode. He proposed that floodwaters had risen over continents already in their present positions, buried them for a short period only—"a universal and transient deluge," in his words—and left as their memorial only a superficial layer of loam and gravel, and a set of topographical features carved by the waters as they rose and fell.

Reliquiae diluvianae is not a waffling, grandiose theoretical treatise on all effects and causes of the Flood, but a specific empirical study of caves and their associated fauna. Buckland had previously examined a cave at Kirkdale, in Yorkshire, and had won the Royal Society's Copley Medal for his efforts. Now he expanded his work to other caves in Britain and to a series of caverns and fissures in Germany.

As his general argument for the importance of caves in attesting to a recent and transient flood, Buckland held that the rising waters had so disturbed all open-air environments that only secluded caves preserved fair evidence for the integrity of antediluvian communities.

So completely has the violence of that tremendous convulsion destroyed and remodeled the form of the antediluvian surface, that it is only in caverns, that have been protected from its ravages that we may hope to find undisturbed evidence of events in the period immediately preceding it.

The caves were full of bones, trapped within by the rising waters. The bones belonged to species then resident in their local areas (thus the Flood was not sufficiently violent

An interesting perspective on early nineteenth-century styles of excavation. A cave containing various "relics of the flood" yields its treasures. FROM BUCKLAND, 1823.

to mix faunas in a random hodgepodge throughout the world). The bones were fresh (pointing to a recent burial), enclosed only with mud washed in by floodwaters or by a light covering of cave drippings (also indicating a deluge of no great antiquity), and belonged to species now extinct but closely allied with modern forms (the less fortunate creatures that found no lodging on the ark).

Buckland's discussion of the Kirkdale cave provides a good illustration of his methods and modes of argument. He found an extensive deposit of fossil bones, broken into angular fragments, sometimes embedded in mud, sometimes encrusted with drippings of cave limestone. Invoking a gastronomical simile from his own time, Buckland described his cache:

Where the mud was shallow, and the heaps of teeth and bones considerable, parts of the latter were elevated some inches above the surface of the mud and its stalagmitic crust; and the upper ends of the bones thus projecting like the legs of pigeons through a piecrust into the void space above, have become thinly covered with stalagmitic drippings, whilst their lower extremities have no such incrustation, and have simply the mud adhering to them in which they have been inbedded.

Buckland devotes most of his monograph to proving that Kirkdale was a hyena den, and that the bones present therein had been gathered and crushed by its denizens. He worked, as all good geologists do, by seeking modern analogues for his ancient results. He learned everything he could about hyenas, ranging from the Latin texts of classical authors to personal observations of hyenas in the Exeter zoo. He proved that the Kirkdale bones were crushed and cracked into the same angular fragments that modern hyenas produce, and he found that the curious spheres of bone fragments within his caves were identical with the droppings of his friends behind bars at Exeter. He also discovered abundant hyena bones within the cave—all crushed and cracked as well—indicating that hyenas approach their own dead as they treat the prey and carrion of other species that form their usual diet.

Since Buckland found no uncracked hyena bones in the cave (but did recover some in outside deposits), he conjectured that as floodwaters rose, the hyenas had left the cave and high-tailed it for the hills:

Should it be further asked, why we do not find, at least, the entire skeleton of the one or more hyenas that died last and left no survivors to devour them; we find a sufficient reply to this question, in the circumstance of the probable destruction of the last individuals by the diluvian waters: on the rise of these, had there been any hyenas in the den, they would have rushed out, and fled for safety to the hills; and if absent, they could by no possibility have returned to it from the higher levels: that they were extirpated by this catastrophe is obvious, from the discovery of their bones in the diluvial gravel both of England and Germany.

The most common bones at Kirkdale belonged to elephants, rhinos, and hyenas. Since all these animals now inhabit tropical climates, Buckland assumed that the deluge had marked a rapid transition to colder temperatures. (He was quite wrong, for we now know that all these species were long-haired, Ice Age variants of modern tropical relatives.) The *Reliquiae diluvianae* is most distinctive in avoiding any discussion of causes and general theories. Buckland abjured the older traditions of system building and speculation, and wrote instead an empirical monograph on specific evidences for a flood. This tactic rendered his work testable and laid the ground for its refutation—the most healthy activity that science can pursue. In discussing the supposed shift to colder climates, Buckland made his only conjecture about cause and then immediately withdrew in conformity with his larger goal:

What this cause was, whether a change in the inclination of the earth's axis, or the near approach of a comet, or any other cause or combination of causes purely astronomical, is a question the discussion of which is foreign to the object of the present memoir.

After discussing Kirkdale and other caves of Britain and Germany, Buckland moved on to subsidiary evidence for a universal deluge. The last part of *Reliquiae diluvianae* discusses two corroborating sources. First, Buckland studied the loams and gravels that mantle solid strata throughout northern Europe, and he found within them bones of the same animals that frequented his caves. Since he regarded loams and gravels as direct deposits of the Flood, similar fossils established the cave remains as relics of the last days before Noah. Secondly, he argued that the sculpturing of hills and valleys records the action of surging floodwaters.

In summarizing his discussion of Kirkdale, Buckland drew an essential inference that sowed the seeds of his later undoing. Buckland's flood theory absolutely required two conclusions to establish Noah's deluge as the agent that both sealed the caves and deposited exterior loams and gravels. First, all cave deposits and gravels must represent material of the same age. Second, each of these accumulations must record a single event, not a series of floods or other catastrophes.

> There is no alternation of this mud with beds of bone or of stalagmite, such as would have occurred had it been produced by land floods often repeated; once, and once only, it appears to have been introduced; and we may consider its vehicle to have been the turbid waters of the same inundation that produced universally the diluvial gravel and loam on the surface without.

In drawing the inference, Buckland had left his self-proclaimed, strictly empirical path (a misplaced ideal that few imaginative scientists can and do follow in any case). No real data supported his claim for contemporaneity of cave deposits with loams and gravels. Moreover, since his caves were widely separated, he could present no direct evidence that the fossils within all hailed from the same time. Indeed, Buckland was arguing in reverse—from prior belief to empirical conclusion. He assumed that these diverse and discontinuous deposits were contemporaneous because he believed so strongly in the historical reality of Noah's flood. Yet, he also claimed that he could prove Noah's flood from the empirical evidence alone. You can't have it both ways.

Nonetheless, in a bold and striking conclusion, penned four years earlier in his inaugural address at Oxford in 1819, Buckland proclaimed:

> The grand fact of an universal deluge at no very remote period is proved on grounds so decisive and

incontrovertible, that had we never heard of such an event from Scripture or any other authority, Geology of itself must have called in the assistance of some such catastrophe.

This famous quotation has often been exposed to ridicule on the assumption that Buckland suffered from advanced self-delusion born of his biblical convictions. Not so. The statement, though forceful, is not unreasonable and reflects one of the supreme ironies in all the history of science.

We know, in retrospect, that England and most of northern Europe were, quite recently, covered several times by massive continental ice sheets. The evidences that glaciers leave—large boulders transported far from their source, poorly sorted gravels apparently dumped into their present resting place by catastrophic agents—are very similar to what gigantic floods might produce. Indeed, much glacial topography is formed by floodwaters from melting ice. Buckland, in fact, was studying evidence of glaciation but, quite naturally, interpreted his data as results of flooding. If Buckland had lived in southern Europe or if the science of geology had arisen in the tropics, this reasonable version of "flood theory" would never have entered our history. We can scarcely blame Buckland for not envisaging a mile of ice atop his native land. Surely, in the 1820s, the idea of a continental ice sheet was preposterous and unthinkable, while a surging flood confuted neither reason nor experience. However, and again in retrospect, we can easily see why Buckland's theory quickly failed the test. He attributed his cave deposits and external gravels to a *single flood;* they were, in fact, produced by *several* episodes of *glaciation.*

Throughout the 1820s, Buckland's theory was a subject of lively debate within the Geological Society of London. The greatest geologists of Britain lined up on opposite sides. As his chief ally, Buckland could depend upon his Cambridge counterpart and fellow divine, the Reverend Adam Sedgwick. Leading the opposition were Charles Lyell, the great apostle of gradualism, and the aristocratic Roderick Impey Murchison. The debate surged with all the vigor of Buckland's floodwaters, but within ten years both

Buckland and Sedgwick had thrown in the towel.

Two primary discoveries forced Buckland's retreat. First, he eventually had to admit that his deposits of loam and gravel were not distributed throughout the world (as "an universal deluge" would require) but only over lands in northern latitudes (reflecting—though Buckland did not yet know the reason—the limited extent of glaciers spreading from polar regions).

Second, and more importantly, the everyday dog work of geology proved that Buckland's caves and gravels did not all correlate, or "match up," as products of a single event in time and also that several deposits recorded more than one episode of flooding (or glaciation, as we would now say). "Correlation" is the basic activity of field geologists. We walk from outcrop to outcrop; we try to trace the beds of one location to the strata of another; we ascertain which beds at our first location match (or correlate in time with) sets of strata in other places.

As this basic work proceeded, geologists recognized that Buckland's cave deposits and gravels represented many events, not a single universal flood. This discovery did not require that floods be abandoned as causal agents, but it did rob Noah of any special status. If numerous floods had occurred, then Buckland's striking evidence could not be ascribed to any particular biblical event. Moreover, since Buckland found no human bones in any of his deposits (whereas Noah's deluge occurred to extirpate rapacious humanity), he eventually concluded that *all* the many floods he now recognized had antedated the Noachian deluge.

In 1829, following a vigorous debate at the Geological Society over Conybeare's paper on the Thames Valley (William Conybeare was a prominent member of Buckland's team), Lyell wrote triumphantly to his supporter Gideon Mantell:

Murchinson and I fought stoutly and Buckland was very piano. Conybeare's memoir is not strong by any means. He admits three deluges before the Noachian! and Buckland adds God knows how many catastrophes besides, so we have driven them out of the Mosaic record fairly.

(For nonmusical readers, I point out that piano simply means "soft" in Italian. The instrument bears its name as a shortening of pianoforte for a device that can play both softly, or piano, and loudly, or forte.)

Buckland himself admitted defeat on the same grounds in his next major book of 1836, though he had not yet recognized the glacial alternative:

> Discoveries which have been made, since the publication of this work [*Reliquiae diluvianae*], show that many of the animals therein described, existed during more than one geological period preceding the catastrophe by which they were extirpated. Hence it seems more probable, that the event in question, was the last of the many geological revolutions that have been produced by violent irruptions of water, rather than the comparatively tranquil inundation described in the Inspired Narrative.

When their evidence fails, fine scientists like Buckland do not simply admit defeat, crawl into a hole, and don a hair shirt. They retain their interest and struggle to find new explanations. Buckland not only abandoned his flood theory when empirical work disproved it, but he eventually led the movement in Britain to substitute ice for water.

Although study in retrospect is unfair to historical figures, I must report that I experienced an almost eerie feeling while reading *Reliquiae diluvianae* in the light of later knowledge about glacial theory. So many of Buckland's specific empirical statements almost cry out for interpretation by ice sheets rather than water. He continually reports, for example, that the inundation, both in Britain and in North America, must have come from the north, an obvious direction for advancing ice but not for universal floodwaters of a rising ocean. He also argues that blocks of granite moved to lower altitudes from the summit of Mont Blanc prove that the Flood rose high enough to cover all mountains—while we would simply say that descending glaciers brought the boulders down.

Louis Agassiz, the Swiss geologist who had grown up almost literally between mountain glaciers, developed the theory of ice ages during the 1830s. He and Buckland became fast friends and co-explorers. Buckland also became one of England's first converts to glacial theory. He read three papers advocating this new interpretation of his old evidence before the Geological Society in 1840 and 1841, and he eventually even persuaded his old adversary Charles Lyell about the reality and power of continental ice sheets. Thus, Buckland not only promptly abandoned his flood theory when it failed the test; he also led the search for new explanations and rejoiced in their discovery.

Modern creationists, on the other hand, have dogmatically preached an even more outmoded and discredited version of flood theory since G.M. Price revived it fifty years ago. They do no fieldwork to test their claims (arguing instead by distorting the work of true geologists for rhetorical effect), and they will change not one jot or tittle of their preposterous theory.

I can present no greater contrast between this modern pseudoscience and the truly scientific spirit than Adam Sedgwick's recantation in his presidential address before the Geological Society of London in 1831. As Buckland's chief supporter, he had led the fight for flood theory; but he knew by then that he had been wrong. He also recognized that he had argued poorly at a critical point: he had correlated the caves and gravels not by empirical evidence, but by a prior scriptural belief in the Flood's reality. As empirical evidence disproved his theory, he realized this logical weakness and submitted himself to rigorous self-criticism. I know no finer statement in all the annals of science than Sedgwick's forthright recantation, and I wish to end this essay with his words. As a witness at the Arkansas creationism trial in December 1981, I also read this passage into the courtroom record because I felt that it illustrated so well the difference between dogmatism, which cannot change, and true science, done in this case by people who happened to be creationists. The final irony and deep message is simply this: flood theory, the centerpiece of modern creationism, was disproved 150 years ago, largely by professional clergymen who were also geologists, exemplary scientists, and creationists. The enemy of knowledge and science is irrationalism, not religion:

Having been myself a believer, and, to the best of my power, a propagator of what I now regard as a philosophic heresy, and having more than once been quoted for opinions I do not now maintain, I think it right, as one of my last acts before I quit this Chair, thus publicly to read my recantation. . . .

There is, I think, one great negative conclusion now incontestably established—that the vast masses of diluvial gravel, scattered almost over the surface of the earth, do not belong to one violent and transitory period. . . .

We ought, indeed, to have paused before we first adopted the diluvian theory, and referred all our old superficial gravel to the action of the Mosaic flood. . . . In classing together distant unknown formations under one name; in giving them a simultaneous origin, and in determining their date, not by the organic remains we had discovered, but by those we expected hypothetically hereafter to discover, in them; we have given one more example of the passion with which the mind fastens upon general conclusions, and of the readiness with which it leaves the consideration of unconnected truths.

8 | False Premise, Good Science

MY VOTE for the most arrogant of all scientific titles goes without hesitation to a famous paper written in 1866 by Lord Kelvin, "The 'Doctrine of Uniformity' in Geology Briefly Refuted." In it, Britain's greatest physicist claimed that he had destroyed the foundation of an entire profession not his own. Kelvin wrote:

> The "Doctrine of Uniformity" in Geology, as held by many of the most eminent of British geologists, assumes that the earth's surface and upper crust have been nearly as they are at present in temperature and other physical qualities during millions of millions of years. But the heat which we know, by observation, to be now conducted out of the earth yearly is so great, that if *this* action had been going on with any approach to uniformity for 20,000 million years, the amount of heat lost out of the earth would have been about as much as would heat, by 100° Cent., a quantity of ordinary surface rock of 100 times the earth's bulk. (See calculation appended.) This would be more than enough to melt a mass of surface rock equal in bulk to the *whole earth*. No hypothesis as to chemical action, internal fluidity, effects of pressure at great depth, or possible character of substances in the interior of the earth, possessing the smallest vestige of probability, can justify the supposition that the earth's crust has

remained nearly as it is, while from the whole, or from
any part, of the earth, so great a quantity of heat has
been lost.

I apologize for inflicting so long a quote so early in the
essay, but this is not an extract from Kelvin's paper. It is the
whole thing (minus the appended calculation). In a mere
paragraph, Kelvin felt he had thoroughly undermined the
very basis of his sister discipline.

Kelvin's arrogance was so extreme, and his later comeup-
pance so spectacular, that the tale of his 1866 paper, and of
his entire, relentless forty-year campaign for a young earth,
has become the classical moral homily of our geological
textbooks. But beware of conventional moral homilies.
Their probability of accuracy is about equal to the chance
that George Washington really scaled that silver dollar clear
across the Rappahannock.

The story, as usually told, goes something like this. Geology, for several centuries, had languished under the thrall
of Archbishop Ussher and his biblical chronology of but a
few thousand years for the earth's age. This restriction of
time led to the unscientific doctrine of catastrophism—the
notion that miraculous upheavals and paroxysms must
characterize our earth's history if its entire geological story
must be compressed into the Mosaic chronology. After long
struggle, Hutton and Lyell won the day for science with
their alternative idea of uniformitarianism—the claim that
current rates of change, extrapolated over limitless time,
can explain all our history from a scientific standpoint by
direct observation of present processes and their results.
Uniformity, so the story goes, rests on two propositions:
essentially unlimited time (so that slow processes can
achieve their accumulated effect), and an earth that does not
alter its basic form and style of change throughout this vast
time. Uniformity in geology led to evolution in biology and
the scientific revolution spread. If we deny uniformity, the
homily continues, we undermine science itself and plunge
geology back into its own dark ages.

Yet Kelvin, perhaps unaware, attempted to undo this triumph of scientific geology. Arguing that the earth began as a molten body, and basing his calculation upon loss of heat from the earth's interior (as measured, for example, in mines), Kelvin recognized that the earth's solid surface could not be very old—probably 100 million years, and 400 million at most (although he later revised the estimate downward, possibly to only 20 million years). With so little time to harbor all of evolution—not to mention the physical history of solid rocks—what recourse did geology have except to its discredited idea of catastrophes? Kelvin had plunged geology into an inextricable dilemma while clothing it with all the prestige of quantitative physics, queen of the sciences. One popular geological textbook writes (C.W. Barnes, in bibliography), for example:

> Geologic time, freed from the constraints of literal biblical interpretation, had become unlimited; the concepts of uniform change first suggested by Hutton now embraced the concept of the origin and evolution of life. Kelvin single-handedly destroyed, for a time, uniformitarian and evolutionary thought. Geologic time was still restricted because the laws of physics bound as tightly as biblical literalism ever had.

Fortunately for a scientific geology, Kelvin's argument rested on a false premise—the assumption that the earth's current heat is a residue of its original molten state and not a quantity constantly renewed. For if the earth continues to generate heat, then the current rate of loss cannot be used to infer an ancient condition. In fact, unbeknown to Kelvin, most of the earth's internal heat is newly generated by the process of radioactive decay. However elegant his calculations, they were based on a false premise, and Kelvin's argument collapsed with the discovery of radioactivity early in our century. Geologists should have trusted their own intuitions from the start and not bowed before the false lure of physics. In any case, uniformity finally won and scientific geology was restored. This transient episode teaches us that we must trust the careful empirical data of a profession and not rely too heavily on theoretical interventions from outside, whatever their apparent credentials.

So much for the heroic mythology. The actual story is by no means so simple or as easily given an evident moral interpretation. First of all, Kelvin's arguments, although fatally flawed as outlined above, were neither so coarse nor as unacceptable to geologists as the usual story goes. Most geologists were inclined to treat them as a genuine reform of their profession until Kelvin got carried away with further restrictions upon his original estimate of 100 million years. Darwin's strong opposition was a personal campaign based on his own extreme gradualism, not a consensus. Both Wallace and Huxley accepted Kelvin's age and pronounced it consonant with evolution. Secondly, Kelvin's reform did not plunge geology into an unscientific past, but presented instead a different *scientific* account based on another concept of history that may be more valid than the strict uniformitarianism preached by Lyell. Uniformitarianism, as advocated by Lyell, was a specific and restrictive theory of history, not (as often misunderstood) a general account of how science must operate. Kelvin had attacked a legitimate target.

KELVIN'S ARGUMENTS AND THE REACTION OF GEOLOGISTS

As codiscoverer of the second law of thermodynamics, Lord Kelvin based his arguments for the earth's minimum age on the dissipation of the solar system's original energy as heat. He advanced three distinct claims and tried to form a single quantitative estimate for the earth's age by seeking agreement among them (see Joe D. Burchfield's *Lord Kelvin and the Age of the Earth,* the source for most of the technical information reported here).

Kelvin based his first argument on the age of the sun. He imagined that the sun had formed through the falling together of smaller meteoric masses. As these meteors were drawn together by their mutual gravitational attraction, their potential energy was transformed into kinetic energy, which, upon collision, was finally converted into heat, causing the sun to shine. Kelvin felt that he could calculate the total potential energy in a mass of meteors equal to the sun's bulk and, from this, obtain an estimate of the sun's original heat. From this estimate, he could calculate a minimum age for the sun, assuming that it has been shining at its present intensity since the beginning. But this calculation was crucially dependent on a set of factors that Kelvin could not really estimate—including the original number of meteors and their original distance from each other—and he never ventured a precise figure for the sun's age. He settled on a number between 100 and 500 million years as a best estimate, probably closer to the younger age.

Kelvin based his second argument on the probable age of the earth's solid crust. He assumed that the earth had cooled from an originally molten state and that the heat now issuing from its mines recorded the same process of cooling that had caused the crust to solidify. If he could measure the rate of heat loss from the earth's interior, he could reason back to a time when the earth must have contained enough heat to keep its globe entirely molten—assuming that this rate of dissipation had not changed through time. (This is the argument for his "brief" refutation of uniformity, cited at the beginning of this essay.) This argument sounds more "solid" than the first claim based on a hypothesis about how the sun formed. At least one can hope to measure directly its primary ingredient—the earth's current loss of heat. But Kelvin's second argument still depends upon several crucial and unprovable assumptions about the earth's composition. To make his calculation work, Kelvin had to treat the earth as a body of virtually uniform composition that had solidified from the center outward and had been, at the time its crust formed, a solid sphere of similar temperature throughout. These restrictions also prevented Kelvin from assigning a definite age for the solidification of the earth's crust. He ventured between 100 and 400 million years, again with a stated preference for the smaller figure.

Kelvin based his third argument on the earth's shape as a spheroid flattened at the poles. He felt that he could relate this degree of polar shortening to the speed of the earth's rotation when it formed in a molten state amenable to flattening. Now we know—and Kelvin knew also—that the earth's rotation has been slowing down continually as a result of tidal friction. The earth rotated more rapidly when it first formed. Its current shape should therefore indicate its age. If the earth formed a long time ago, when rotation was quite rapid, it should now be very flat. If the earth is not so ancient, then it formed at a rate of rotation not so different from its current pace, and flattening should be less. Kelvin felt that the small degree of actual flattening indicated a relatively young age for the earth. Again, and for the third time, Kelvin based his argument upon so many unprovable assumptions (about the earth's uniform composition, for example) that he could not calculate a precise figure for the earth's age.

Thus, although all three arguments had a quantitative patina, none was precise. All depended upon simplifying assumptions that Kelvin could not justify. All therefore yielded only vague estimates with large margins of error. During most of Kelvin's forty-year campaign, he usually cited a figure of 100 million years for the earth's age—plenty of time, as it turned out, to satisfy nearly all geologists and biologists.

Darwin's strenuous opposition to Kelvin is well recorded, and later commentators have assumed that he spoke for a troubled consensus. In fact, Darwin's antipathy to Kelvin was idiosyncratic and based on the strong personal commitment to gradualism so characteristic of his world view. So wedded was Darwin to the virtual necessity of unlimited time as a prerequisite for evolution by natural selection that he invited readers to abandon *The Origin of Species* if they could not accept this premise: "He who can read Sir Charles Lyell's grand work on the *Principles of Geology,* and yet does not admit how incomprehensively vast have been the past periods of time, may at once close this volume." Here Darwin commits a fallacy of reasoning—the confusion of gradu-

sm with natural selection—that characterized all his work
and that inspired Huxley's major criticism of the *Origin:*
"You load yourself with an unnecessary difficulty in adopt-
ing *Natura non facit saltum* [Nature does not proceed by
leaps] so unreservedly." Still, Darwin cannot be entirely
blamed, for Kelvin made the same error in arguing explicitly that his young age for the earth cast grave doubt upon
natural selection as an evolutionary mechanism (while not
arguing against evolution itself). Kelvin wrote:

> The limitations of geological periods, imposed by
> physical science, cannot, of course, disprove the hy-
> pothesis of transmutation of species; but it does seem
> sufficient to disprove the doctrine that transmutation
> has taken place through "descent with modification by
> natural selection."

Thus, Darwin continued to regard Kelvin's calculation of
the earth's age as perhaps the gravest objection to his theory. He wrote to Wallace in 1869 that "Thomson's [Lord
Kelvin's] views on the recent age of the world have been for
some time one of my sorest troubles." And, in 1871, in
striking metaphor, "But then comes Sir W. Thomson like
an odious spectre." Although Darwin generally stuck to his
guns and felt in his heart of hearts that something must be
wrong with Kelvin's calculations, he did finally compromise
in the last edition of the *Origin* (1872), writing that more
rapid changes on the early earth would have accelerated the
pace of evolution, perhaps permitting all the changes we
observe in Kelvin's limited time:

> It is, however, probable, as Sir William Thompson [*sic*]
> insists, that the world at a very early period was sub-
> jected to more rapid and violent changes in its physical
> conditions than those now occurring; and such
> changes would have tended to induce changes at a
> corresponding rate in the organisms which then ex-
> isted.

Darwin's distress was not shared by his two leading supporters in England, Wallace and Huxley. Wallace did not tie
the action of natural selection to Darwin's glacially slow
time scale; he simply argued that if Kelvin limited the earth
to 100 million years, then natural selection must operate at
generally higher rates than we had previously imagined. "It
is within that time [Kelvin's 100 million years], therefore,
that the whole series of geological changes, the origin and
development of all forms of life, must be compressed." In
1870, Wallace even proclaimed his happiness with a time
scale of but 24 million years since the inception of our fossil
record in the Cambrian explosion.

Huxley was even less troubled, especially since he had
long argued that evolution might occur by saltation, as well
as by slow natural selection. Huxley maintained that our
conviction about the slothfulness of evolutionary change
had been based on false and circular logic in the first place.
We have no independent evidence for regarding evolution
as slow; this impression was only an inference based on the
assumed vast duration of fossil strata. If Kelvin now tells us
that these strata were deposited in far less time, then our
estimate of evolutionary rate must be revised correspond-
ingly.

> Biology takes her time from geology. The only reason
> we have for believing in the slow rate of the change in
> living forms is the fact that they persist through a
> series of deposits which, geology informs us, have
> taken a long while to make. If the geological clock is
> wrong, all the naturalist will have to do is to modify his
> notions of the rapidity of change accordingly.

Britain's leading geologists tended to follow Wallace and
Huxley rather than Darwin. They stated that Kelvin had
performed a service for geology in challenging the virtual
eternity of Lyell's world and in "restraining the reckless
drafts" that geologists so rashly make on the "bank of
time," in T.C. Chamberlin's apt metaphor. Only late in his

campaign, when Kelvin began to restrict his estimate from a vague and comfortable 100 million years (or perhaps a good deal more) to a more rigidly circumscribed 20 million years or so did geologists finally rebel. A. Geikie, who had been a staunch supporter of Kelvin, then wrote:

> Geologists have not been slow to admit that they were in error in assuming that they had an eternity of past time for the evolution of the earth's history. They have frankly acknowledged the validity of the physical arguments which go to place more or less definite limits to the antiquity of the earth. They were, on the whole, disposed to acquiesce in the allowance of 100 millions of years granted them by Lord Kelvin, for the transaction of the long cycles of geological history. But the physicists have been insatiable and inexorable. As remorseless as Lear's daughters, they have cut down their grant of years by successive slices, until some of them have brought the number to something less than ten millions. In vain have geologists protested that there must be somewhere a flaw in a line of argument which tends to results so entirely at variance with the strong evidence for a higher antiquity.

KELVIN'S SCIENTIFIC CHALLENGE AND THE MULTIPLE MEANINGS OF UNIFORMITY

As a master of rhetoric, Charles Lyell did charge that anyone who challenged his uniformity might herald a reaction that would send geology back to its prescientific age of catastrophes. One meaning of uniformity did uphold the integrity of science in this sense—the claim that nature's laws are constant in space and time, and that miraculous intervention to suspend these laws cannot be permitted as an agent of geological change. But uniformity, in this methodological meaning, was no longer an issue in Kelvin's time, or even (at least in scientific circles) when Lyell first published his *Principles of Geology* in 1830. The scientific catastrophists (see essay 7) were not miracle mongers, but men who fully accepted the uniformity of natural law and sought to render earth history as a tale of *natural* calamities occurring infrequently on an ancient earth.

But uniformity also had a more restricted, substantive meaning for Lyell. He also used the term for a particular theory of earth history based on two questionable postulates: first, that rates of change did not vary much throughout time and that slow and current processes could therefore account for all geological phenomena in their accumulated impact; second, that the earth had always been about the same, and that its history had no direction, but represented a steady state of dynamically constant conditions.

Lyell, probably unconsciously, then performed a clever and invalid trick of argument. Uniformity had two distinct meanings—a methodological postulate about uniform laws, which all scientists had to accept in order to practice their profession, and a substantive claim of dubious validity about the actual history of the earth. By calling them both uniformity, and by showing that all scientists were uniformitarians in the first sense, Lyell also cleverly implied that, to be a scientist, one had to accept uniformity in its substantive meaning as well. Thus, the myth developed that any opposition to uniformity could only be a rearguard action against science itself—and the impression arose that if Kelvin was attacking the "doctrine of uniformity" in geology, he must represent the forces of reaction.

In fact, Kelvin fully accepted the uniformity of law and even based his calculations about heat loss upon it. He directed his attack against uniformity only upon the substantive (and dubious) side of Lyell's vision. Kelvin advanced two complaints about this substantive meaning of uniformity. First, on the question of rates. If the earth were substantially younger than Lyell and the strict uniformitarians believed, then modern, slow rates of change would not be sufficient to render its history. Early in its history, when the earth was hotter, causes must have been more energetic and intense. (This is the "compromise" position that Darwin finally adopted to explain faster rates of change early in the history of life.) Second, on the question of direction. If the earth began as a molten sphere and lost heat continually through time, then its history had a definite pattern and path of change. The earth had not been perennially the same, merely changing the position of its lands and seas in a never-ending dance leading nowhere. Its history followed a definite road, from a hot, energetic sphere to a cold, listless world that, eventually, would sustain life no longer. Kelvin fought, within a scientific context, for a *short-term, directional* history against Lyell's vision of an essentially eternal steady-state. Our current view represents the triumph of neither vision, but a creative synthesis of both. Kelvin was both as right and as wrong as Lyell.

RADIOACTIVITY AND KELVIN'S DOWNFALL

Kelvin was surely correct in labeling as extreme Lyell's vision of an earth in steady-state, going nowhere over untold ages. Yet, our modern time scale stands closer to Lyell's concept of no appreciable limit than to Kelvin's 100 million years and its consequent constraint on rates of change. The earth is 4.5 billion years old.

Lyell won this round of a complicated battle because Kelvin's argument contained a fatal flaw. In this respect, the story as conventionally told has validity. Kelvin's argument was not an inevitable and mathematically necessary set of claims. It rested upon a crucial and untested assumption that underlay all Kelvin's calculations. Kelvin's figures for heat loss could measure the earth's age only if that heat represented an original quantity gradually dissipated through time—a clock ticking at a steady rate from its initial reservoir until its final exhaustion. But suppose that new heat is constantly created and that its current radiation from the earth reflects no original quantity, but a modern process of generation. Heat then ceases to be a gauge of age.

Kelvin recognized the contingent nature of his calculations, but the physics of his day included no force capable of generating new heat, and he therefore felt secure in his assumption. Early in his campaign, in calculating the sun's age, he had admitted his crucial dependence upon no new source of energy, for he had declared his results valid "unless new sources now unknown to us are prepared in the great storehouse of creation."

Then, in 1903, Pierre Curie announced that radium salts constantly release newly generated heat. The unknown source had been discovered. Early students of radioactivity quickly recognized that most of the earth's heat must be continually generated by radioactive decay, not merely dissipating from an originally molten condition—and they realized that Kelvin's argument had collapsed. In 1904, Ernest Rutherford gave this account of a lecture, given in Lord Kelvin's presence, and heralding the downfall of Kelvin's forty-year campaign for a young earth:

> I came into the room, which was half dark, and presently spotted Lord Kelvin in the audience and realized that I was in for trouble at the last part of the speech dealing with the age of the earth, where my views conflicted with his. To my relief, Kelvin fell fast asleep, but as I came to the important point, I saw the old bird sit up, open an eye and cock a baleful glance at me! Then a sudden inspiration came, and I said Lord Kelvin had limited the age of the earth, provided no new source of heat was discovered. That prophetic utterance refers to what we are now considering tonight, radium!

Thus, Kelvin lived into the new age of radioactivity. He never admitted his error or published any retraction, but he privately conceded that the discovery of radium had invalidated some of his assumptions.

The discovery of radioactivity highlights a delicious double irony. Not only did radioactivity supply a new source of heat that destroyed Kelvin's argument; it also provided the clock that could then measure the earth's age and proclaim it ancient after all! For radioactive atoms decay at a constant rate, and their dissipation does measure the duration of time. Less than ten years after the discovery of radium's newly generated heat, the first calculations for radioactive decay were already giving ages in billions of years for some of the earth's oldest rocks.

We sometimes suppose that the history of science is a simple story of progress, proceeding inexorably by objective accumulation of better and better data. Such a view underlies the moral homilies that build our usual account of the advance of science—for Kelvin, in this context, clearly impeded progress with a false assumption. We should not be beguiled by such comforting and inadequate stories. Kelvin proceeded by using the best science of his day, and colleagues accepted his calculations. We cannot blame him for not knowing that a new source of heat would be discovered. The framework of his time included no such force. Just as Maupertuis lacked a proper metaphor for recognizing that embryos might contain coded instructions rather than preformed parts (see next essay), Kelvin's physics contained no context for a new source of heat.

The progress of science requires more than new data; it needs novel frameworks and contexts. And where do these fundamentally new views of the world arise? They are not simply discovered by pure observation; they require new modes of thought. And where can we find them, if old modes do not even include the right metaphors? The nature of true genius must lie in the elusive capacity to construct these new modes from apparent darkness. The basic chanciness and unpredictability of science must also reside in the inherent difficulty of such a task.

9 | For Want of a Metaphor

IN 1745, the great French savant Pierre-Louis Moreau de Maupertuis wrote a little book with a big theme and an odd title. (The original measures but 5½ by 3¼ inches and includes fewer than 200 pages of text, printed, thanks to the ample margins of a more generous age, in the even smaller space of 3¼ by 1¾ inches.) He called it *Vénus physique*—the "physical," or "earthly, Venus," or, more loosely, "physical love" (as opposed to the interpretive, spiritual, or psychological dimensions of this subject of the centuries). It presents, as the title implies, a wide-ranging account of the natural history of procreation—a primer in how various animals do it. We learn, for example, in juxtaposed contrast that

> the impetuous bull, proud of his strength, does not amuse himself with caresses; he throws himself immediately upon the heifer; he penetrates deeply into her loins and squirts there, in large streams, the liquid that must make her fertile. The turtle dove, by tender calls, announces his love: a thousand caresses, a thousand pleasures precede the last pleasure.

Proceeding down the scale of being (as his century conceived it), Maupertuis reaches the hermaphroditic land snails and discusses their darts. (Many land snails develop a calcareous "arrow" with a beautifully formed tip. In the

139

elaborate rituals that precede copulation, the snail acting as male thrusts its dart repeatedly into its partner's muscular foot. The dart is not part of the penis, and beyond the obvious observation that it serves some role in sexual stimulation, we still do not know its precise function.) Maupertuis didn't have an answer either, but he made an interesting, if fatuous, analogy:

> What is the function of this organ? Perhaps this animal, so cold and so slow in all its operations, needs to be excited by these stings. Men made cold by age, or whose senses have become enfeebled, sometimes have recourse to equally violent means in order to awake in them the passions of love. Oh unhappy man, who tries to excite by pain the feelings that should only arise from voluptuousness! . . . Oh innocent snail, you are perhaps the only creature for whom these means are not criminal—because they are for you an effect of Nature's order. Receive then, and render, a thousand times the stings of these darts that arm you.

At the bottom of the scale, Maupertuis encountered a special problem with hydras, the soft-bodied, freshwater relatives of corals. Maupertuis and his colleagues considered hydras as transitional forms between plants and animals because they reproduce either by budding new individuals off a parental stalk or by regenerating entire bodies from disarticulated fragments of the same stalk. Maupertuis, in no uncertain terms, had identified pleasure as nature's end in the process of reproduction:

> Nature has the same interest in perpetuating all species: she has inspired in each the same theme, and that theme is pleasure. It is pleasure that, in the human species, drives everything before it—that, despite a thousand obstacles opposed to the union of two hearts, a thousand torments that must follow, conducts lovers towards the purpose that nature has ordained.

But if pleasure be nature's order, then how can the lowly hydra enjoy reproduction by having its stalk cut into pieces?

> What is one to think of this strange style of reproduction, of this principle of life extended into each piece of the animal. . . . In other animals, nature has attached pleasure to the act that multiplies them; could it be that nature has endowed this creature with some sort of voluptuous feeling when it is cut into pieces?

Perhaps these passages inspired Maupertuis's decision to publish anonymously, although he lived in a century so refreshingly less prudish than the one that followed (while his direct and charming words also stand in such favorable contrast to the perpetual, self-conscious analysis of our own age). Nonetheless, *Vénus physique* is not primarily a work about the natural history of love, whatever the value of these sections for publicity and immediate renown. It is, for most of its length, a sophisticated treatise on the science of embryology—on the most direct and enduring physical effects of love. The title, perhaps, was a come-on, but the book is a masterpiece.

Maupertuis was born in France in 1698. Although he ranged widely across the disciplinary boundaries imposed by a later age, he won his reputation for work in the physical sciences—both for his courage in introducing and expounding Newton's work in a nation strongly wedded to Descartes's alternatives and for directing an arduous expedition to Lapland that affirmed Newton's prediction of an earth not perfectly spherical, but flattened at the poles. This combination of care and daring won him Voltaire's support and his star rose. In 1738, Voltaire recommended to Frederick the Great that Maupertuis might be just the man to direct his rehabilitated Academy of Sciences in Berlin. Maupertuis took the job and flourished in it for several years. But a series of involved intrigues brought him down and incurred Voltaire's undying wrath and the deadly satire of his acerbic pen. Maupertuis was eventually exonerated, but he never recovered his health and reputation, and he died, a broken man, in 1759.

Like many general treatises, the *Vénus physique* had its origin in a specific problem. In a culture with deep racist traditions, human skin color has exerted a perpetual fascination, and no aspect of the subject inspired more interest than the occasional discovery of peculiar individuals who seemed to breach the boundaries. Jeremiah's God, pessimistic about redemption among those who had fallen by the wayside, proclaimed: "Can the Ethiopian change his skin, or the leopard his spots?" But some humans did transgress beyond the limits of apparently stable categories, leading to fear that one's own future relatives might stray or that the categories themselves might not be so comfortably fixed in their conventional statuses of relative worth. Essay 22 discusses a Caucasian woman with large patches of melanic skin, who fascinated a London physician in 1813. But her case was rare and irrelevant. A more general phenomenon was, however, fairly common and thus both threatening and fascinating—namely, albinism among black people. Albinism is well known among most, or all, dark-skinned vertebrate races and species; albino blacks, paler than any Caucasian, are not rare, and the trait is inherited in family lines.

An albino child of black parents had been on exhibit in Paris and Maupertuis's thoughts and observations served as the inspiration for his *Vénus physique*. His work bears the subtitle: *Dissertation physique à l'occasion du nègre blanc* ("A physical dissertation inspired by the white Negro"). *Vénus physique* contains two parts, the much longer first section on embryology and the natural history of love and a forty-five-page closing statement on the origin of human races. (This second section contains some poorly formed evolutionary speculations and is largely responsible for Maupertuis's reputation as a Darwinian precursor—an unfair and anachronistic assessment, based on a few fleeting passages that abstract Maupertuis from the concerns of his time. *Vénus physique* is a treatise on embryology and the exciting debates of his own century.)

This second section features a discussion of human biogeography and tries to explain a false pattern reconstructed from unreliable reports by travelers—a belief that blacks inhabited the tropics, while arctic regions were the exclusive preserve of giants and dwarfs. Maupertuis argues, in short, that superior white races had simply pushed all miscreants and oddballs out of the more favorable temperate regions. We can easily see how the albino child had inspired Maupertuis's thoughts for this second section, but what influence could he have exerted over the heart of *Vénus physique*—the first, long section on embryology? An answer to this question provides the key to *Vénus physique* and a proper assessment of Maupertuis's creative and unusual opinion in the great embryological debate of his age.

In one of the hottest arguments of eighteenth-century science, students of development lined up on both sides of an ancient dichotomy stretching back to Greek science. Aristotle had argued that embryonic development is both the greatest of all biological mysteries and the key to a deep understanding of organisms—propositions that remain as true today (for our ignorance is still profound) as when the "master of them that know" proclaimed them more than two thousand years ago. Greek scientists had envisaged two broad types of solutions, and their eighteenth-century successors continued to respect the categories. One group, the preformationists, argued that embryology must represent an unfolding of preexisting structure. A tiny homunculus must be curled up in the egg or sperm. It need not be a perfect miniature of the adult—for the relative form and position of parts may change with growth—but the structures must all be present and connected from the first. A second group including Maupertuis, the epigeneticists, argued that the visual appearance of development must be respected as a literal truth. The embryo seems to differentiate complex parts from an original simplicity, and so it must be in reality (preformationists, in response, claimed that contemporary microscopes were too poor to see preformed parts in the tiny and gelatinous early embryo). Embryology is addition and differentiation, not mere unfolding.

We must reject the silly good guy–bad guy scenario usually attached in false retrospect to this tale: namely, that preformationists were blinded by theological prejudice against change of any sort and therefore imposed upon the egg or sperm what they could not observe—while epigeneticists were heroes of empirical science and merely called it as they saw it under their microscopes.

In fact, the preformationists maintained an idea of science far closer to our own. They were the mechanists who insisted upon a material cause for all phenomena. And they were stuck in the limited knowledge of their own century. What alternative did they have? The wondrous complexity of a human body cannot develop mysteriously from an original formless nothingness; organs must therefore be present from the start. Most epigeneticists, on the other hand, were comfortable with a view of causality that we would now reject as "vitalistic"—the idea that an external, nonmaterial force could impose complex design upon a fertilized egg that began with unformed potential alone.

Maupertuis was a conspicuous oddball in this great debate, for he was both a fervent epigeneticist and a committed mechanist. Unlike his epigeneticist colleagues, therefore, he expected to find material precursors for all parts in eggs and sperm. But these parts could not form a prebuilt homunculus. They must be totally scattered and completely disaggregated. They must also exist in numbers far in excess of what the embryo needs (for if eggs and sperm included all the right parts, and these only, then Maupertuis might have been labeled an odd preformationist who advocated a disarticulated homunculus). Embryonic development must therefore represent a selection, sorting out, attraction, and creative joining of these separated parts, not a mere enlargement of structures already fixed in form, place, or number. But how could disaggregated parts come together, and how could the right ones be sorted out and joined (or the wrong ones occasionally incorporated in abnormal fetuses)? The idea of a preformed homunculus seemed to present fewer problems.

Several of Maupertuis's arguments against preformationism were the conventional retorts of his age. Against the ovists (those who placed the homunculus in the female egg) Maupertuis raised the usual, and always troubling, issue of encapsulation. The egg cells of the homunculus must contain other, vastly smaller, homunculi and so on back to countless generations of inconceivable tininess. All of human history, in fact, must have been prefigured in the ovaries of Eve.

> Eggs destined to produce males each contain only a single male. But an egg with a female contains not only that female, but also her ovaries, in which other females, already fully formed, are enclosed—the source of infinite generation. Can matter be divisible to infinitude; can the form of a fetus that will be born in a thousand years be as distinct as the one that will be born in nine months?

And why then do males exist at all? Does their semen merely release and inspire the previously lifeless homunculus? Was this, Maupertuis asks, the fire that Prometheus stole from the gods?

Against the spermaticists (those who placed the homunculi within sperm cells), Maupertuis raised the additional problem that many million cells are expelled in each ejaculation. Could nature be so profligate and endow millions of unused cells with perfect homunculi that will never enjoy life?

> This little worm, swimming in the seminal fluid, contains an infinity of generations, from father to father. And each [homunculus] has his seminal fluid, full of swimming animals so much smaller than himself. . . . And what prodigiousness when we consider the number and tiny size of these animals. One man calculated that a single pike fish, in one generation, could produce more pikes than there are men on earth, even assuming that all the earth is as densely populated as Holland. . . . Such an immense richness, such a fecundity without limit in nature; do we not have here a

prodigality of resources! May we not say that the expense and outlay are excessive!

Maupertuis ventured an alternate, and interestingly incorrect, function for the recently discovered "spermatick animalcules," as his English colleagues called them. He imagined that they stirred and mixed the seminal fluids of male and female, thus bringing together the parts that must form the embryo.

But Maupertuis added to this great embryological contretemps some new arguments and a strikingly original perspective. For centuries, the turf of debate had been the embryo and its observable process of development. True creativity often resides in the joining of previously disparate fields, in the recognition that apparently unrelated phenomena from other disciplines may offer solutions to old dilemmas. And so, finally, we return to albinos and to Maupertuis's creative insight.

Maupertuis was among the first European scientists to trace the pedigrees of unusual traits through family lines. He recognized that these results, apparently so unrelated to embryology, might solve the great debate in favor of epigenesis. He compiled a pedigree for polydactyly (extra fingers) through three generations of a German family and proved one cardinal fact, long touted in anecdote and folklore but never conclusively established: inheritance passes through *both* male and female lines; that is, extra fingers could be inherited from either fathers or mothers. Maupertuis then recognized that this feature of heredity rather than any direct aspect of embryology, might resolve the problem of development; for how could preformationism be supported if *both* parents can contribute to the form of their offspring. If homunculi are curled up in eggs or sperm, the noncontributing parent should not play an equal role in the form of its offspring. And what about hybrids who bear characteristic features of two parents belonging to different species? Maupertuis concluded:

It seems to me that one of these systems [either the ovist or spermaticist version of preformationism] is completely destroyed by the resemblance of the child, sometimes to the father, and sometimes to the mother, and by the intermediate animals born from parents of two different species. . . . Since the child resembles both, I believe we must conclude that both parents play an equal role in its development.

Maupertuis could not compute the pedigree of the albino child of black parents on exhibit in Paris. But he learned that albinism had been traced through family lines among blacks in Senegal, and he argued that albinism would, like polydactyly, drive another nail into the preformationist coffin. In any case, the "white Negro" inspired him to organize his thoughts on development and to write one of the classics of eighteenth-century science.

Preformationists had a standard response to the phenomenon of biparental inheritance. They argued that one parent carried the homunculus, while the seminal fluid of the second parent modified it. Maupertuis ridiculed this argument, especially as applied to the development of mules from a horse and a donkey parent:

If the fetus were in the [spermatic] worm that swims in the seminal fluid of its father, why should it sometimes resemble its mother? If it were only in the egg of its mother, what form would it have in common with its father? If the little horse were already formed in the mare's egg, would it develop a donkey's ears because [the seminal fluid of] an ass puts parts of the egg into motion?

Despite Maupertuis's rhetoric, the preformationist response was not absurd when applied to ordinary, continuously varying traits. The short homunculus of one parent might be lengthened by spermatic fluid from a taller parent; even the ears of a horse might be elongated after contact with the vigorous motions of an ass's semen. But Maupertuis had a very strong argument for odd and discrete traits —polydactyly and albinism, for example. These features looked exactly the same in offspring, whether they were inherited from fathers or mothers. Could one really believe that inheritance would work precisely the same way for traits rigidly preformed in a homunculus and those merely caused by motions in the seminal fluid of the noncontributing parent?

Maupertuis also used the white Negro to formulate two other arguments against preformationism. First, he considered albinism as a kind of deformity and therefore analogized it with monstrous births (from Siamese twinning to polydactyly). Such fetal abnormalities posed a great problem for preformationists. If they accepted the monstrosity as fully preformed, they faced the theological dilemma of a bumbling or malevolent deity who would plan such unhappiness and program it into Eve's ovaries. If they argued (as they usually did) that extra parts indicated the accidental amalgamation of two homunculi, then did it not stretch the imagination to claim that a polydactylous child received nearly all of itself from one germ and merely an extra finger from another? Such an explanation wouldn't serve for albinism in any case. One had to believe that a few white homunculi had been scattered into the progenitor of black people. Possible but unlikely.

Second, Maupertuis argued that preformationism could not easily explain the origin of different human skin colors from a single progenitor:

> The first mother must have contained eggs of different colors, which themselves carried an innumerable suite of eggs for the same color . . . but which would only hatch in the time that Providence had marked for the origin of the people contained in them. It would not be impossible that, one day, the suite of white eggs for peopling our region would run out, and all European nations would change color; as it would not be impossible also that the source of black eggs might become exhausted, and that Ethiopia would only have white inhabitants.

Maupertuis effectively used the white Negro to marshal his arguments against preformationism, but what could he offer to explain his own peculiar brand of epigenesis? Since he refused to admit any vitalistic, external directing forces, he had to discover a source of order within the seminal fluids themselves. How did all the disaggregated parts come together and why did the right ones usually join—thus making it so difficult for an extra part to insinuate itself and accounting for the rarity of such abnormalities as polydactyly? Here Maupertuis reached an impasse, though he labored mightily to break through.

His best suggestion harked back to the Newtonian perspective so important to his general scientific view. If gravity regulated the attraction of physical objects, then a kind of gravity must bring the right parts together to form a fetus. Eye parts would have a natural affinity for nose parts, nose for teeth, and so forth, until a whole animal might be constructed, just as the dry bones came to life on Ezekiel's desert. Moreover, maternal and paternal eye parts would have an equal chance to join the fetus and a completed embryo would therefore be an amalgam of traits from both parents.

> Why, if this force [gravity in the sense of attraction] exists in nature, should it not regulate the formation of animal bodies? If the seminal fluid of each parent contains parts destined to form the heart, the head, the gut, the arms, the legs, and if each of these parts has a greater affinity for union with neighboring parts of the completed animal than with any other part, then the fetus will build itself.

Maupertuis felt that monsters with extra parts offered special support for his gravitational theory because supernumerary organs always formed in the right place. An extra finger never protrudes from the belly or the back of the head, but always joins the other five, thus proving that finger parts have a natural affinity for each other and for neighboring sections of the hand.

Since I love to participate in intellectual battles, if only vicariously, I was most excited, in reading the *Vénus physique,* by the opportunity to watch a brilliant man struggling to explain the greatest mystery of biology and knowing full well, despite all the effort strung out over two hundred pages, that he had not succeeded. Maupertuis sensed that his gravitational theory was weak, based on no direct evidence, and rooted more in analogy than in any concrete observation. Yet, he had to propose something and could think of nothing better. For Maupertuis was firmly committed both to his general mechanistic perspective and to the specific theory of epigenesis—and these intellectual positions forced him to argue that material particles to form the fetus must exist in the seminal fluids of both parents, for vitalistic forces could not direct the differentiation of complex structures from nothing. He opted for disaggregated parts, mixed together in the seminal fluid but able somehow to find each other and form the embryo. And he correctly sensed the unsatisfactory and improbable nature of his theory.

We would say today that Maupertuis's basic insight was correct: complexity cannot arise from formless potential; something must exist in the egg and sperm. But we now hold a radically different concept of this "something." Where Maupertuis could not think beyond actual parts, we have discovered programmed instructions. Eggs and sperm do not carry parts themselves, but only coded instructions, written in DNA, to direct the building of a proper embryo.

But how could Maupertuis have reached this elegant solution, for his century lacked analogs in thought and technology to imagine a process of abstraction from actual parts to programmed rules for their construction. Programmed instructions were not part of the intellectual equipment of eighteenth-century thinkers. Music boxes pointed in the right direction, but the first revolutionary invention based on programmed instructions, the Jacquard loom, was not introduced until the early 1800s. This automatic weaving device, with instructions on punched cards, directly inspired Hollerith's later invention of data cards for census machines (later transmogrified to the famous IBM computer card—do not fold, spindle, or mutilate). How could Maupertuis imagine the correct solution to his dilemma—

programmed instructions—in a century that had no player pianos, not to mention computer programs?

We often think, naïvely, that missing data are the primary impediments to intellectual progress—just find the right facts and all problems will dissipate. But barriers are often deeper and more abstract in thought. We must have access to the right metaphor, not only to the requisite information. Revolutionary thinkers are not, primarily, gatherers of facts, but weavers of new intellectual structures. Ultimately, Maupertuis failed because his age had not yet developed a dominant metaphor of our own time—coded instructions as the precursor to material complexity.

3 | The Importance of Taxonomy

10 | Of Wasps and WASPs

"HE IS HURLING the insult of the century against our mothers, wives, daughters and sisters, under the pretext of making a great contribution to scientific research." Thus did Louis B. Heller, congressman from New York, label the Kinsey report on *Sexual Behavior in the Human Female* (1953) in a letter to the postmaster general, urging that the book be banned from the mails. Dr. Henry Van Dusen, president of the Union Theological Seminary, doubted Kinsey's facts but proclaimed that if true nonetheless, "they reveal a prevailing degradation in American morality approximating the worst decadence of the Roman empire." "The most disturbing thing," Van Dusen continued in castigating Kinsey's report, "is the absence of a spontaneous, ethical revulsion from the premises of the study."

Yet the premises seemed uncomplicated enough. Kinsey had sought, through extensive interviews with more than 5,000 women, to compile a statistical record of what people do do, rather than what law and custom say they should do. He passed no judgment and merely reported his findings; he did, however, discover a frequency of premarital and extramarital sexual relations that, to say the least, disturbed the chivalric code of many naïve, hypocritical, or smugly satisfied people—particularly older men in power.

Alfred C. Kinsey suffered the misfortune of publishing his report in 1953 at the height of McCarthyite hysteria in

America (his earlier 1948 report on *Sexual Behavior in the Human Male* had caused a stir but had not inspired such calumny, perhaps because society has always accepted a wider range of behavior among males and because the early political climate of post-war years had been much more liberal). Many labeled Kinsey's report on female sexuality as an exercise in communism or, if not directly subversive, sufficiently weakening of American moral fiber to allow easy communist access to our troubled shores. A special House Committee, established to investigate the use of funds by tax-exempt foundations and led by noted cold warrior B. Carroll Reece, dragged the Rockefeller Foundation onto its carpet. The foundation capitulated to these and other pressures, and Kinsey's main source of support ended abruptly in 1954. The Reece Committee issued its majority report in December 1954, accusing some foundations of using tax-exempt monies for studies "directly supporting subversion." The Kinsey reports were explicitly cited as unworthy of the aid they had received. Kinsey never did find an alternate source of support; he died two years later, overworked, angry, and distressed that so many years of further data might never see publication (renewed funding arrived later, but not in time for Kinsey's personal vindication).

Kinsey was no lifelong crusader for sexual enlightenment. He drifted into sex research almost by accident (though not without prior interest). He had been trained as an entomologist and was, at the time of his shift in careers, one of America's foremost taxonomists of wasps (six-legged, not two-legged). Soon after his switch, he began a Phi Beta Kappa lecture at Indiana University with these words:

> With individual variation as a biologic phenomenon I have been concerned during some twenty years of field exploration and laboratory research. In the intensive and extensive measurement of tens of thousands of small insects which you have probably never seen, and about which you certainly cannot care, I have made some attempt to secure the specific data and the

quantity of data on which scientific scholarship must be based. During the past two years, as a result of a convergence of circumstances, I have found myself confronted with material on variation in certain types of human behavior.

Most people, when they learn about Kinsey's earlier career, tend to regard the discovery with quaint amusement. How odd that a man who later shook America should have spent most of his professional career on the taxonomy of tiny insects. Surely there can be no relationship between two such disparate careers. As one wag wrote in a graffito on the title page to Harvard's only copy of Kinsey's greatest monograph on wasps: "Why don't you write about something more interesting, Al?"

I wish to argue, however, that Kinsey's wasps and WASPs were intimately related by his common intellectual approach to both. And since wasps preceded WASPs, Kinsey's career as a taxonomist had a direct and profound impact upon his sex research. In fact, Kinsey pursued his sex research by following a particular "taxonomic way of thought," a valid style of science that does not match most stereotypes of the enterprise. The special character of Kinsey's work—the aspects that brought him such fame and trouble—flowed directly from the taxonomic approach he had learned and perfected as an entomologist.

Aside from the specific conclusions that so shocked America—basically the high frequency of things that nice people supposedly didn't do, from homosexuality to premarital and extramarital sex among women to the high frequency of sexual contact with animals among men who had grown up on farms—Kinsey stirred the world with his different procedural approach to sex research. He worked with three basic premises, all flowing directly from his taxonomic perspective. First, he would base his conclusions upon samples far larger than any previous researcher had gathered. No more extrapolations to all of humanity from a small and homogeneous population of college students. Second, his sample would be heterogeneous—old and young, farm and city, poor and rich, illiterate and college educated. As wasps varied from tree to tree, classes, sexes, and generations might differ widely in their sexual behavior. Third, he would pass no judgments but merely describe what people did.

Kinsey received his Ph.D. in entomology from Harvard and then accepted a post as assistant professor of zoology at the University of Indiana, where he remained all his life. He spent the first twenty years of his career in a study, conducted with unprecedented detail, of the taxonomy, evolution, and biogeography of gall-forming wasps in the genus *Cynips*. These small wasps lay their eggs within the tissues of plants (usually the leaves or stems of oaks). When the larvae hatch, they induce the plant to form a gall about them, thus securing both protection and a source of food. The larvae mature within their galls, eventually emerging as winged insects to begin the process anew. Kinsey presented his work on *Cynips* in a number of shorter papers and two large monographs, *The Gall Wasp Genus* Cynips: *A Study in the Origin of Species* (1930) and *The Origin of Higher Categories in* Cynips (1936)—see bibliography.

In 1938, in response to student petition, the university established a noncredit course on marriage (a euphemism, I suppose, for some sex education). Kinsey, who had planned to spend the rest of his life studying wasps, was asked to serve as chairman of the committee to regulate this course and to give three lectures on the biology of sex. Kinsey was conscientious and empirically oriented to a fault. He went to the library to find the required information about human sexual response—and he couldn't. So he decided that he would have to compile it himself. He began by interviewing students but soon realized that he was not getting representative information about American heterogeneity. He began to travel on weekends, gathering information in nearby towns at his own expense. He developed an extensive format for interviews and wrote the responses in code to assure anonymity (Kinsey's intuitive skill as an interviewer became legendary). He recorded enormous variation in sexual behavior among people of different economic status, extending his researches to Gary, Chicago, Saint Louis, and to Indiana prisons. As his work became more public, criticism mounted, but the university remained firm in its support of Kinsey's right to know.

Eventually, with the university's backing, he established the Institute for Sex Research and secured Rockefeller Foundation money for his burgeoning interviews and their publication. His work culminated in two great volumes, *Sexual Behavior in the Human Male* and *Sexual Behavior in the Human Female,* each based on more than 5,000 interviews with white Americans of diversified backgrounds. (True to his convictions about the fundamental character of variability, Kinsey knew that he did not have enough data to reach conclusions about black Americans or to extrapolate to other nations and cultures.) Long before these volumes appeared, Kinsey had, with great reluctance and sadness but with creeping inevitability, abandoned the wasp studies that had brought him so much pleasure and had set his standards of scientific work.

Although Kinsey confined his major works on wasps to a single family, the Cynipidae, his aims were as broad as natural history itself. He thought deeply about the practice and meaning of classification and hoped to reformulate the principles of taxonomy. He wrote in 1927:

> From our work on Cynipidae, in connection with a study of the published work in other fields of taxonomy, I propose to attempt a formulation of the philosophy of taxonomy, its usefulness as a means of portraying and explaining species as they exist in nature, and its importance in the coordination and elucidation of biologic data.

Kinsey felt that he could achieve these larger goals by performing a specific study with such unprecedented factual detail that larger principles would emerge from the volume of information itself. Kinsey was a workaholic before the word was invented. On a traveling fellowship in 1919–1920, he logged 18,000 miles (2,500 on foot) in southern and western regions of the United States and collected some 300,000 specimens of gall wasps. His two trips to rural Mexico and Central America in the 1930s were monuments to his insatiable industry. Still, he was never satisfied. In his 1936 monograph, he lamented that for each of his 165 species, he had collected, on average, "only" 214 insects and 755 galls. For 51 of these species (variable groups in regions of uniform topography), he stated that he would not be satisfied until he had gathered a grand total of 1,530,000 insects and 3 to 4 million galls!

More than mere collection mania underlay Kinsey's expressed desires and actual efforts. A modern statistician might well argue that Kinsey had an inadequate appreciation of sampling theory; you really don't need to get every one. Still, Kinsey pursued his copious collecting because he operated and centered his biological beliefs upon one cardinal principle: the primacy and irreducibility of variation.

Ironically, much of taxonomic practice had not fully assimilated this fundamental change brought to biology by evolutionary theory. Many taxonomists still viewed the world as a series of pigeonholes, each housing a species. Species, in this view, should be defined by their "essences" —fundamental features separating them from all others. Variation was regarded as a nuisance at best—a kind of accidental splaying out around the essential form, and serving only to create confusion in the correct assignment of pigeonholes. Most classical taxonomists treated variation as a necessary evil and often established species after studying only a few specimens.

Taxonomists like Kinsey, who understood the full implications of evolutionary theory, developed a radically different attitude to variation. Islands of form exist, to be sure: cats do not flow together in a sea of continuity, but rather come to us as lions, tigers, lynxes, tabbies, and so forth. Still, although species may be discrete, they have no immutable essence. Variation is the raw material of evolutionary change. It represents the fundamental reality of nature, not an accident about a created norm. Variation is primary; essences are illusory. Species must be defined as ranges of irreducible variation.

This antiessentialist way of thinking has profound consequences for our basic view of reality. Ever since Plato cast shadows on the cave wall, essentialism has dominated Western thought, encouraging us to neglect continua and to divide reality into a set of correct and unchanging categories. Essentialism establishes criteria for judgment and worth: individual objects that lie close to their essence are good; those that depart are bad, if not unreal.

Antiessentialist thinking forces us to view the world differently. We must accept shadings and continua as fundamental. We lose criteria for judgment by comparison to some ideal: short people, retarded people, people of other beliefs, colors, and religions are people of full status. The taxonomic essentialist scoops up a handful of fossil snails in a single species, tries to abstract an essence, and rates his snails by their match to this average. The antiessentialist sees something entirely different in his hand—a range of irreducible variation defining the species, some variants more frequent than others, but all perfectly good snails. Ernst Mayr, our leading taxonomic theorist, has written elegantly and at length on the difference between essentialism and variation as an ultimate reality ("population thinking" in his terminology—see his recent book, *The Growth of Biological Thought*).

Kinsey, who understood the implications of evolutionary theory so well, was a radical antiessentialist in taxonomy. His belief in the primacy of variation spurred an almost frantic effort to collect ever more specimens. His belief in continua forced him to explore virtually every square foot of suitable territory for *Cynips* in North America—for whenever he found large gaps, he strongly suspected (usually correctly) that intermediate forms would be found in some geographically contiguous area.

In the end, Kinsey's antiessentialism became almost too radical. He was so convinced that species would grade into other species that he began to name truly intermediate geographical variants *within* a single species as separate entities, and established a bloated taxonomy of full names for transient and minor local variants. (Kinsey decided that species arose by the spread through local populations of discrete mutations with small effects. Thus, whenever he found a local population differing from others by mutations of the sort produced in laboratory stocks, he established a new species. But local populations within a species often establish small mutations without losing their central tie to the rest of the species—the ability to interbreed.)

More important for American social history, Kinsey transported bodily to his sex research the radical antiessentialism of his entomological studies. Kinsey's twenty years with *Cynips* may not be judged as a wasteful diversion compared with the later source of his fame. Rather, Kinsey's wasp work established both the methodology and principles of reasoning that made him a pioneer in sex research.

I am not merely making learned inferences about continuities that the master of antiessentialism didn't recognize. Kinsey knew perfectly well what he was doing. He regretted not a moment spent on wasps, both because he loved them too, and because their study had set his intellectual sights. In the first chapter of his first treatise on *Sexual Behavior in the Human Male*, Kinsey included a remarkable section on "the taxonomic approach," with two subheadings—"in biology," followed by the explicit transfer, "in applied and social sciences." Kinsey wrote:

> The techniques of this research have been taxonomic, in the sense in which modern biologists employ the term. It was born out of the senior author's long-time experience with a problem in insect taxonomy. The transfer from insect to human material is not illogical, for it has been a transfer of a method that may be applied to the study of any variable population.

Extensive sampling was the hallmark of Kinsey's work. Most earlier studies of human sexual behavior had either confined their reporting to unusual cases (Krafft-Ebing's *Psychopathia Sexualis,* for example) or had generalized from small and homogeneous samples. If Kinsey had hoped for millions of wasps and their galls, he would at least interview many thousands of humans. He knew that he needed such large numbers because his antiessentialist perspective proclaimed two truths about variation for wasps and people alike—apparently homogeneous populations in one place (all college students at Indiana or all murderers at Alcatraz) would exhibit an enormous range of irreducible variation, and discrete local populations in different places (older middle-class women in Illinois or poor young men in New York) would differ greatly in average sexual behaviors. (Biologists refer to these two types of variation as within-population and between-population.) Kinsey decided that he would have to sample many differing groups and large numbers within each group. He wrote in the first paragraph of his treatise on males:

> It is a fact-finding survey in which an attempt is made to discover what people do sexually, and what factors account for differences in sexual behavior among individuals, and among various segments of the population.

In his section on "the taxonomic approach in biology" he explained why his experience with wasps had set his methods for humans:

> Modern taxonomy is the product of an increasing awareness among biologists of the uniqueness of individuals, and of the wide range of variation which may occur in any population of individuals. The taxonomist is, therefore, primarily concerned with the measurement of variation in series of individuals which stand as representatives of the species in which he is interested.

Kinsey's belief in the primacy of variation and diversity became a crusade. His 1939 Phi Beta Kappa lecture, "Individuals," focused on the "unlimited nonidentity" among organisms in any population and castigated both biological and social scientists for drawing general conclusions from small and relatively homogeneous samples. For example:

A mouse in a maze, today, is taken as a sample of all individuals, of all species of mice under all sorts of conditions, yesterday, today, and tomorrow. A half dozen dogs, pedigrees unknown and breeds unnamed, are reported on as "dogs"—meaning all kinds of dogs—if, indeed, the conclusions are not explicitly or at least implicitly applied to you, to your cousins, and to all other kinds and descriptions of humans. . . . A noted American colloid chemist startles the country with the announcement of a new cure for drug addicts; and it is not until other laboratories report failure to obtain similar results that we learn that the original experiments were based on a half dozen individuals.

As a second important transfer from his entomologically based antiessentialism, Kinsey repeatedly emphasized the impossibility of pigeonholing human sexual response by allocating people into rigidly defined categories. As his wasps formed chains of continuity from one species to the next, human sexual response could be fluid, changing, and devoid of sharp boundaries. Of male homosexuality, he wrote:

Males do not represent two discrete populations, heterosexual and homosexual. The world is not to be divided into sheep and goats. Not all things are black nor all things white. It is a fundamental of taxonomy that nature rarely deals with discrete categories. Only the human mind invents categories and tries to force facts into separate pigeon-holes. The living world is a continuum in each and every one of its aspects. The sooner we learn this concerning human sexual behavior the sooner we shall reach a sound understanding of the realities of sex.

The third transfer—the one that ultimately brought Kinsey so much trouble—raised the contentious issue of judgment. If variation is primary, copious, and irreducible, and if species have no essences, then what "natural" criterion can we discover for judgment? An odd variant is as much a member of its species as an average individual. Even if average individuals are more common than peculiar organisms, who can identify one or the other as "better"—for species have no "right" form defined by an immutable essence. Kinsey wrote in "Individuals," again making explicit reference to wasps:

> Prescriptions are merely public confessions of prescriptionists. . . . What is right for one individual may be wrong for the next; and what is sin and abomination to one may be a worthwhile part of the next individual's life. The range of individual variation, in any particular case, is usually much greater than is generally understood. Some of the structural characters in my insects vary as much as twelve hundred percent. This means that populations from a single locality may contain individuals with wings 15 units in length, and other individuals with wings 175 units in length. In some of the morphologic and physiologic characters which are basic to the human behavior which I am studying, the variation is a good twelve thousand percent. And yet social forms and moral codes are prescribed as though all individuals were identical; and we pass judgments, make awards, and heap penalties without regard to the diverse difficulties involved when such different people face uniform demands.

Kinsey often claimed in his two great reports that he had merely recorded the facts of sexual behavior without either passing or even implying judgment. On the prefatory page to his report on males, he wrote:

> For some time now there has been an increasing awareness among many people of the desirability of obtaining data about sex which would represent an accumulation of scientific fact completely divorced from questions of moral value and social custom.

His critics countered by arguing that an absence of judgment in the context of such extensive recording is, itself, a form of judgment. I think I would have to agree. I see no possibility for a completely "value-free" social science. Kinsey may have disclaimed in the reports themselves, but the statement just quoted from his 1939 essay makes no bones about his conviction that nonjudgmental attitudes are morally preferable—and his basic belief in the primacy of variation has evident implications itself. Can one despise what nature provides as fundamental? (One can, of course, but few people will favor an ethic that rejects life and the world as we inevitably find them.)

What, in any case, is the alternative? Should we not compile the factual data of human sexual behavior? Or should people who undertake such a study sprinkle each finding with an irrelevant assessment of its moral worth from their personal point of view? That would be hubris indeed. Ultimately, however, I must confess that my approval of Kinsey, and my strong attraction to him, arises from our shared values. I too am a taxonomist.

At the beginning of *The Grapes of Wrath,* as Tom Joad heads home after a prison term, he meets Casy, his old preacher. Casy explains that he no longer holds revivals because he could not reconcile his own sexual behavior (often inspired by the fervor of the revival meeting itself) with the content of his preaching:

> I says, "Maybe it ain't a sin. Maybe it's just the way folks is." . . . Well, I was layin' under a tree when I figured that out, and I went to sleep. And it come night, an' it was dark when I come to. They was a coyote squawkin' near by. Before I knowed it, I was sayin' out loud . . . "There ain't no sin and there ain't no virtue. There's just stuff people do. . . . And some of the things folks do is nice, and some ain't nice, but that's as far as any man got a right to say."

11 | Opus 100*

THROUGHOUT A LONG DECADE of essays I have never, and for definite reasons, written about the biological subject closest to me. Yet for this, my hundredth effort, I ask your indulgence and foist upon you the Bahamian land snail *Cerion,* mainstay of my own personal research and field-work. I love *Cerion* with all my heart and intellect but have consciously avoided it in this forum because the line between general interest and personal passion cannot be drawn from a perspective of total immersion—the image of doting parents driving friends and neighbors to somnolent distraction with family movies comes too easily to mind. These essays must follow two unbreakable rules: I never lie to you, and I strive mightily not to bore you. But, for this one time in a hundred, I will risk the second for personal pleasure alone.

Cerion is the most prominent land snail of West Indian islands. It ranges from the Florida Keys to the small islands of Aruba, Bonaire, and Curaçao, just off the Venezuelan coast, but the vast majority of species inhabit two principal centers—Cuba and the Bahamas. *Cerion*'s life includes little excitement by our standards. Most species inhabit rocks and

---

*This is the fourth volume of essays compiled from my monthly column in *Natural History* magazine. I marked ten years of work, and never a deadline missed (I won't tell you about the numerous close calls), with this act of self-indulgence as my treat to myself for the one hundredth effort.

sparse vegetation abutting the seashore. They may live for five to ten years, but they spend most of this time in the warm weather equivalent of hibernation (called estivation), hanging upside down from vegetation or affixed to rocks. After a rain or sometimes in the relative cool and damp of night, they descend from their twigs and stones, nibble at the fungi on decaying vegetation, and perhaps even copulate. We have marked and mapped the movement of individual snails and many can be found on the same few square yards of turf, year after year.

Why pick *Cerion*? Why, indeed, spend so much time on any detailed particular when all the giddy generalities of evolutionary theory beg for study in a lifetime too short to manage but a few? Iconoclast that I am, I would not abandon the central wisdom of natural history from its inception —that concepts without percepts are empty (as Kant said), and that no scientist can develop an adequate "feel" for nature (that undefinable prerequisite of true understanding) without probing deeply into minute empirical details of some well-chosen group of organisms. Thus, Aristotle dissected squids and proclaimed the world's eternity, while Darwin wrote four volumes on barnacles and one on the origin of species. America's greatest evolutionists and natural historians, G.G. Simpson, T. Dobzhansky, and E. Mayr, began their careers as, respectively, leading experts on Mesozoic mammals, ladybird beetles, and the birds of New Guinea.

Scientists don't immerse themselves in particulars only for the grandiose (or self-serving) reason that such studies may lead to important generalities. We do it for fun. The pure joy of discovery transcends import. And we do it for adventure and for expansion. As drama, Bahamian field trips may seem risible compared with Darwin on the *Beagle*, Bates on the Amazon, and Wallace in the Malay Archipelago—although I would not care to repeat my only close brush with death, caught in a shoot-out among drug runners on North Andros. So much more do I value the quiet times in different worlds: an evening's discussion of bush medicine on Mayaguana, an exploration of ornamental carvings that adorn roofs on Long Island and South Andros, and the finest meal I have ever eaten—a campfire pot of fresh conch stewed with sweet potatoes from Jimmy Nixon's garden on Inagua, after a hot and hard day's work.

If all good naturalists must choose a group of organisms for detailed immersion, we do not select mindlessly or randomly (or even, as some cynics have suggested, because the Bahamas beat the Yukon as a field area). I am interested primarily in the evolution of form and have concentrated on how the varying shapes of an individual's growth can serve as a source of evolutionary change (see my technical book, *Ontogeny and Phylogeny,* in bibliography). An invertebrate paleontologist with these interests would naturally be led to snails, since their shells preserve a complete record of growth from egg to adult.

A student of form with a penchant for gastropods could not avoid *Cerion,* for this genus exhibits, among its several hundred species, a range of form unmatched by any other group of snails. Some *Cerions* are tall and pencil thin; others are shaped like golf balls. When a colleague ventured "square snails" as an example of impossible animals at a public meeting, I was able to show him the peculiar quadrate *Cerion* from the photo on p. 170, bottom row, second from left. Five years ago, I discovered the largest *Cerion,* a thin and parallel-sided fossil giant from Mayaguana more than 70 mm tall. The smallest is a virtual sphere, scarcely 5 mm in diameter, from Little Inagua (see photo).

Cerion's mystery and special interest do not lie in its exuberant diversity alone; many groups of animals include some members with unusual propensities for speciation and consequent variation of form. Species are the fundamental units of biological diversity, distinct populations permanently isolated one from the other by an absence of interbreeding in nature. We should not be surprised that groups producing large numbers of species may become quite diverse in form, since more distinct units provide more opportunities for evolving a wide range of morphologies.

Faced with such a riotous array of shapes, older naturalists

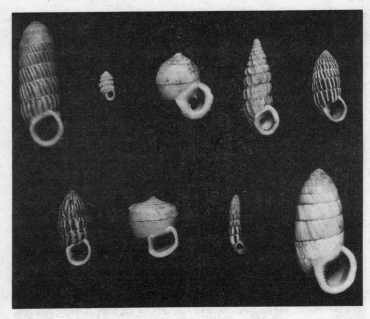

Various *Cerion* shells from the Bahamas and Cuba to show the unparalleled diversity of form within this genus. PHOTO BY AL COLE-MAN.

did name species aplenty in *Cerion,* some 600 of them. But few are biologically valid as distinct noninterbreeding populations. In ten years of fieldwork on all major Bahamian islands, we have only once found two distinct *Cerion* populations living in the same place and not interbreeding—true species, therefore. These included a giant and a dwarf—thus recalling various bad jokes about Chihuahuas and Great Danes. In all other cases, two forms, no matter how distinct in size and shape, interbreed and produce hybrids at their point of geographic contact. Somehow, *Cerion* manages to generate its unparalleled diversity of form without parceling its populations into true species. How can this happen? Moreover, if such different forms hybridize so readily, then the genetic differences between them cannot be great. How

The largest and smallest known *Cerion* shells (and I mean specimens, not representatives of species). I found the giant *Cerion*
excelsior on Mayaguana. The dwarf *C. baconi* lives on Little Inagua.
I estimate the giant's height (missing top restored) at over 70 mm.
PHOTO BY RON ENG.

can such diversity of size and shape arise in the absence of extensive genetic change?

In a related and second mystery, distinct forms of *Cerion* often inhabit widely separated islands. The simplest explanation would propose that these far-flung colonies represent the same species and that hurricanes can blow snails great distances, producing haphazard distributions, or that colonies once inhabiting intermediate islands have become extinct, leaving large distances between survivors. Yet, all *Cerion* experts have developed the feeling (which I share) that these separated colonies, despite their detailed similarity for long lists of traits, have evolved independently *in situ*. If this unconventional interpretation is correct, how can such complex suites of associated traits evolve again and again?

Cerion thus presents two outstanding peculiarities amidst its unparalleled diversity: Its most distinct forms interbreed and are not true species, while these same forms, for all their complexity, may have evolved several times independently. Any scientist who can explain these odd phenomena for *Cerion* will make an important contribution to the understanding of form and its evolution in general. I shall try to describe the few preliminary and faltering steps we have made towards such a resolution.

Cerion has attracted the attention of several prominent naturalists, from Linnaeus, who named its first species in 1758, to Ernst Mayr, who pioneered the study of natural populations 200 years later. Still, despite the efforts of a tiny group of aficionados, *Cerion* has not received the renown it deserves in the light of its curious biology and its promise as an exemplar for the evolution of form. Its relative obscurity can be traced directly to past biological practice. Older naturalists buried *Cerion*'s unusual biology under such an impenetrable thicket of names (for invalid species) that colleagues interested in evolutionary theory have been unable to recover the pattern and interest from utter chaos.

The worst offender was C.J. Maynard, a fine amateur biologist who named hundreds of *Cerion* species from the 1880s through the 1920s. He imagined that he was performing a great service, proclaiming in 1889:

Conchologists may take exception to some of my new species, thinking, perhaps, that I have used too trivial characters in separating them. Believing, however, as I do, that it is the imperative duty of naturalists today, to record minute points of differences among animals . . . I have not hesitated so to designate them, if for no other reason than for the benefit of coming generations.

I trust that I shall not be accused of undue cynicism in recognizing another reason. Maynard financed his Bahamian trips by selling shells, and more species meant more items to flog. *Caveat emptor.*

Professional colleagues were harsh on Maynard's overly fine splitting. H.A. Pilsbry, America's greatest conchologist, declared in uncharacteristically forceful prose that "gods and men may well stand aghast at the naming of individual colonies from every sisal field and potato patch in the Bahamas." W.H. Dall labeled Maynard's efforts as "noxious and stupefying." Yet, when tested in the crucible of practice, neither Pilsbry nor Dall lived up to his brave words. Each recognized at least half the species Maynard advocated, still sufficiently overinflated to bury any pattern in the forest of invalid names.

So rich was *Cerion*'s diversity, and so numerous its species, that G.B. Sowerby, the outstanding English conchologist, who fancied himself (with little justification) a poet, wrote this doggerel in introducing his monograph on the genus:

Things that were not, at thy command,
In perfect form before Thee stand;
And all to their Creator raise
A wondrous harmony of praise.

Sowerby then proceeded to list quite a chorus. And this quatrain dates from 1875, before Maynard ever named a *Cerion!*

In the light of this existing chaos, and before we can even

ask the general questions about form that I posed above, we must pursue a much more basic and humble task. We must find out whether any pattern can be found in the ecological and geographic distribution of *Cerion*'s morphology. If we detect no correlation at all with geography or environment, then what can we explain? Fortunately, in a decade of work, we have reduced the chaos of existing names to predictable patterns and have thereby established the prerequisite for deeper explanation. Of the nature of that deeper explanation, we have intuitions and indications, but neither definite information nor even the tools to provide it (for we are stuck in an area of biology—the genetics of development— that is itself woefully undeveloped). Still, I think we have made a promising start.

I say "we" because I realized right away that I could not do this work alone. I felt competent to analyze the growth and form of shells, but I have no expertise in two areas that must be united with morphology in any comprehensive study: genetics and ecology. So I teamed up with David Woodruff, a biologist from the University of California at San Diego. For a decade we have done everything together, from blisters on Long Island to bullets on Andros.

(I must stop at this point, for I suddenly realize that I have almost broken my first rule. Scientists have a terrible tendency to present their work as a logical package, as if they thought everything out in careful and rigorous planning beforehand and then merely proceeded according to their good designs. It never works that way, if only because anyone who can think and see makes unanticipated discoveries and must fundamentally alter any preconceived strategy. Also, people get into problems for the damnedest of peculiar and accidental reasons. Projects grow like organisms, with serendipity and supple adjustment, not like the foreordained steps of a high school proof in plane geometry. Let me confess. I was first drawn to *Cerion* because I wanted to compare its fossils with snails I had studied on Bermuda. I studiously avoided all modern *Cerions* because I was petrified by the thicket of available names and considered them intractable. Woodruff first went to Inagua because he wanted to study color banding in another genus of snails. But he went at the height of mosquito season and lasted two days. We took our first trip together to Grand Bahama Island: I to study fossils, he to try the other genus again. But I soon discovered that Grand Bahama has no (or very few) rocks of terrestrial origin, hence no fossil land snails. The other genus wasn't much more common. We were stuck there for a week. So we studied the living *Cerions* and found a pattern behind the plethora of names. Since then, following Satchel Paige's advice, we have never looked back.)

About fifteen names had been proposed for the *Cerions* of Grand Bahama and neighboring Abaco Island. After a week, Woodruff and I recognized that only two distinct populations inhabited these islands, each restricted to a definite and different environment.

Abaco and Grand Bahama protrude above a shallow platform called Little Bahama Bank (see accompanying map). When sea level was lower during the last ice age, the entire platform emerged and the islands were connected by land. Little Bahama Bank is separated by deep ocean from the larger Great Bahama Bank, source of the more familiar Bahamian islands (New Providence, with its capital city of Nassau, Bimini, Andros, Eleuthera, Cat, the Exuma chain, and many others). All these islands were also connected during glacial times of low sea level. As Woodruff and I moved from island to island on Great Bahama Bank, we found the same pattern of two different populations, always in the same distinctive environments. On Little Bahama Bank, a dozen invalid names had fallen into this pattern. On Great Bahama Bank, they collapsed, literally by the hundred. About one-third of all *Cerion* "species" (close to 200 in all) turned out to be invalid names based on minor variants within this single pattern. We had reduced a chaos of improper names to a single, ecologically based order. (This reduction applies only to the islands of Little and Great Bahama Bank. Islands on other banks in the southeastern Bahamas, including Long Island, the southeasternmost island of Great Bahama Bank, contain truly different *Cerions*. These *Cerions* can also be reduced to coherent patterns

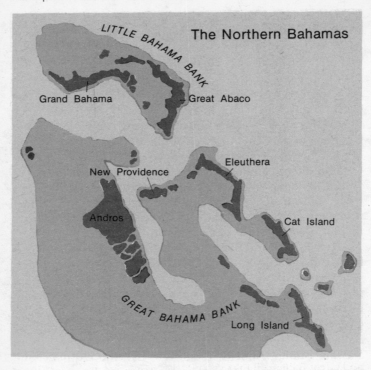

A map of the northern Bahamas, showing edges of the banks.
REPRINTED FROM NATURAL HISTORY.

based on few true species. But one essay can treat just so much, and I confine myself here to the northern Bahamas.)

Bahamian islands have two different kinds of coastlines. Major islands lie at the edge of their banks. The banks themselves are very shallow across their tops but plunge precipitously into deep ocean at their edges. Thus, bank-edge coasts abut the open ocean and tend to be raw and windy. Dunes build along windy coasts and solidify eventually into rock (often mistakenly called "coral" by tourists). Bank-edge coasts are, therefore, usually rocky as well. By contrast, coastlines that border the interior parts of banks —I will call them bank-interior coasts—are surrounded by

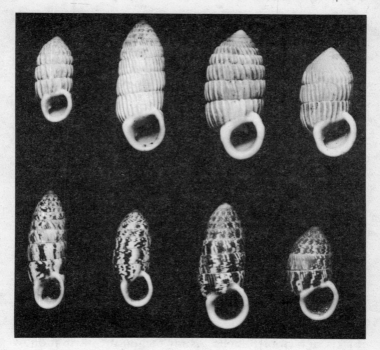

Comparison of ribby, bank-edge *Cerion* (upper row) from Little and Great Bahama Bank, with mottled, bank-interior *Cerion* (lower row). PHOTO BY AL COLEMAN.

calm, shallow waters that extend for miles and do not promote the building of dunes. Bank-interior coasts, therefore, tend to be vegetated, low, and calm.

Woodruff and I found that bank-edge coasts in the northern Bahamas are invariably inhabited by thick-shelled, strongly ribbed, uniformly colored (white to darkish brown), relatively wide, and parallel-sided *Cerions*. To avoid writing most of the rest of this column in Latin, I will skip the formal names and refer to these forms as the "ribby populations" (see photo above). Bank-interior coasts are the home of thin-shelled, ribless or weakly ribbed, variegated (usually with alternating blotches of white and brown), narrow, and barrel-shaped *Cerions*—the "mottled populations." (Mottled *Cerions* also live away from coasts in the centers of islands, while ribby *Cerions* are confined exclusively to bank-edge coasts.)

This pattern is so consistent and invariable that we can "map" hybrid zones even before we visit an island, simply by looking at a chart of bathymetry. Hybrid zones occur where bank-edge coasts meet bank-interior coasts.

This pattern might seem worthy of little more than an indulgent ho-hum. Perhaps mottled and ribby shells are not very different. Maybe the two environments elicit their differing forms directly from the same basic genetic stock, much as good and plentiful food can make a man fat and paltry fare eventually convert the same gent to a scarecrow. The very precision and predictability of the correlation between form and environment might suggest this biologically uninteresting solution. Two arguments, however, seem to stand conclusively against this interpretation and to indicate that mottled and ribby *Cerions* are different biological entities.

First, the ribby snails are not merely mottled forms with thicker and ribbier shells. As my technical contribution to our joint work, I measure each shell in twenty different ways. This effort permits me to characterize both growth and final adult form in mathematical terms. I have been able to show that the differences between ribby and mottled involve several independently varying determinants of form.

Second, an analysis of hybrid zones proves that they mark a mixture of two different entities, not a smooth blending of populations only superficially separate. My morphological analysis shows, in many cases, the anomalies of form, and the increased variation, that so often occur when two different developmental programs are mixed in offspring. Woodruff's genetic analysis also proves that the hybrids combine two substantially different systems, for he finds both generally increased genetic variability in hybrid samples, and genes detected in neither parental population.

We can demonstrate that ribby and mottled represent populations with substantial biological differences, but we cannot specify the cause of separation since we have been

unable to distinguish between two hypotheses. First, ecological: ribby and mottled forms may be recent and immediate adaptations to their differing local environments. White
or light-colored shells are inconspicuous against the bank-
edge background of dune rocks, while thick and ribby shells
protect their bearers on these windy and rocky coasts. Mottled shells are equally inconspicuous (indeed remarkably
camouflaged) when dappled sunlight filters through the
vegetation that houses *Cerion* on most bank-interior coasts,
while thin and light shells are also well suited for hanging
from thin twigs and grass blades. Second, historical: the
pattern may be substantially older (although still probably
adaptive for the reasons cited above). When sea level was
much lower and the banks lay exposed during glacial periods, perhaps ribby populations inhabited all coasts (since
all were then bank-edge), while mottled populations
evolved for life in island interiors. As sea level rose, ribby
and mottled snails simply kept their positions and preferences. The new bank-interior coasts were the interiors of
previously larger islands and they continue as homes of
mottled snails.

The distinction of mottled and ribby resolved nearly all
the two hundred names previously given to *Cerions* from the
northern Bahamas. But one problem (involving about ten
more names) remained. A third kind of *Cerion,* bearing a
thick, but smooth, pure white, and triangularly shaped shell,
had been found on Eleuthera and Cat Island. Previous reports indicated nothing about their ecology or habits, but
we found these thick white snails in two disjunct areas of
southern Eleuthera and in southeastern Cat Island. They
prefer island interiors and fit *Cerion*'s general pattern with
gratifying predictability—that is, they hybridize with mottled populations as we approach bank-interior coasts and
with ribby populations as we move toward bank-edge
coasts. But what are they? Just as ecology and genetics
resolved the basic pattern of mottled and ribby, we must call
upon paleontology to explain our remaining source of diversity.

Fossil dunes of the Bahamas formed at times of high sea
level during warmer periods between episodes of glaciation

(ice ages). Three major sets of dunes built New Providence, the only Bahamian island with a documented geological pedigree (see Garrett and Gould, in bibliography). These include, from youngest to oldest, a few small dunes less than 10,000 years old and deposited since the last glaciers melted; an extensive set (forming the island's backbone), representing the high sea levels of 120,000 years ago, before the last glaciers formed; and a smaller set (situated near the island's center) built more than 200,000 years ago, before a previous glacial period. The oldest dunes contain a fossil *Cerion* now unknown in the Bahamas (see photo on p. 181). The second and most extensive set includes two species of *Cerion,* a dwarf form now extinct and a large, smooth white species called *Cerion agassizi* (named for Alexander Agassiz, son of Louis, and a pioneer of scientific oceanography in the West Indies). The most recent set, as expected, contains either ribby or mottled *Cerions,* as in the modern fauna. We compared the large white snails of Eleuthera and Cat with *C. agassizi* and found no substantial differences. The small populations on these islands are surviving remnants of a species that once lived in abundance on all the islands of Great Bahama Bank.

The two hundred "species" of northern Bahamian *Cerion* therefore reduce to three basic types with a sensible and ordered distribution. Geographic pattern identified ribby and mottled populations, but we needed an assist from history to understand the smooth white shells of Eleuthera and Cat. It is an awfully long stride from this taxonomic exercise in natural history to our ultimate goal—an understanding of how *Cerion*'s unparalleled diversity of form evolved—but we have taken the first step along the only pathway I know.

As an example of how this pattern illuminates the larger question, we have used our distinction of mottled and ribby to prove, for the first time, that the unconventional hypothesis advanced by most *Cerion* experts is indeed valid: the complex suite of characters defining such basic forms as mottled and ribby can evolve independently many times. We find the same distinction of mottled and ribby on both

Fossils from New Providence Island include (from left to right) an extinct form from the lowest dunes; an extinct dwarf (*Cerion universe*) and the large *Cerion agassizi* from the 120,000-year-old dunes; and a ribby snail from the youngest dunes. Remnant populations of *Cerion agassizi* survive on Eleuthera and Cat Island. PHOTO BY AL COLEMAN.

Little and Great Bahama banks. Conventional wisdom would hold that the mottled snails of both banks represent one stock, while the ribby snails of both banks form a different genealogical group. But Daniel Chung, a student of Woodruff's, and Simon Tillier, a leading anatomist of land snails at the Paris Museum, have studied the genital anatomy of these snails for us, and have made the following surprising discovery: *both* mottled and ribby snails of Little Bahama Bank share the *same* anatomy, while both mottled and ribby on Great Bahama Bank share a distinctly different set of genital structures. (Genital anatomy is the standard tool for establishing genealogical affinities among land snails. The differences are sufficiently profound and complex to indicate that shared anatomy reflects common descent while shared shell morphology must evolve independently.) Thus, the complex of traits defining mottled and ribby can evolve again and again. We would not have been able to reach this conclusion had we not extracted the pattern of mottled and ribby from a previous chaos of names.

At this point, I think we begin to peer through a glass darkly at the deeper mystery of form. We have shown that a complex set of independent traits can evolve in virtually the same way more than once. I do not see how this can happen if each trait must develop separately, following its own genetic pathway, each time. The traits must somehow be coordinated in *Cerion*'s genetic program; they must be evoked, or "called forth," together. Some genetic releaser must coordinate the joint appearance of these characters. Does the master genetic program of all *Cerions* encode alternate pathways representing the several basic forms that evolve again and again? The homeotic mutations of insects (see essay 15 in *Hen's Teeth and Horse's Toes*) indicate that some such hierarchical system must regulate development, for the production of well-formed parts in the wrong places (legs where antennae should be, for example) indicates that some master switch must regulate all the genes that produce legs, and that higher controls can turn the master switch on in the wrong place or at the wrong time. Likewise, some master switch within *Cerion*'s program might evoke any one of its basic developmental pathways and evolve the set of traits that marks its fundamental forms again and again.

In this way, *Cerion* provides insight into what may be the most difficult and important problem in evolutionary theory: How can new and complex forms (not merely single features of obvious adaptive benefit) arise if each requires thousands of separate changes, and if intermediate stages make little sense as functioning organisms? If genetic and developmental programs are organized hierarchically, as homeotic mutations and the multiple evolution of basic forms in *Cerion* suggest, then new designs need not arise piecemeal (with all the intractable problems attached to such a view), but in a coordinated way by manipulating the master switches (or "regulators") of developmental programs. Yet so deep is our present ignorance about the nature of development and embryology that we must look at final products (an adult *Drosophila* with a leg where its antenna should be or mottled *Cerions* evolved several times) to make uncertain inferences about underlying mechanisms.

I chose *Cerion* because I thought it might illuminate these large and woolly issues. Yet, although they lurk always in the back of my mind, they are not the source of my daily joy. Little predictions affirmed or small guesses proved wrong and exchanged for more interesting ideas are the food of continual satisfaction. *Cerion,* or any good field project, provides unending stimulation, so long as little puzzles remain as intensely absorbing, fascinating, and frustrating as big questions. Fieldwork is not like the one-hundred-thousandth essay on Shakespeare's sonnets; it always presents something truly new, not a gloss on previous commentaries.

I remember when we found the first population of living *Cerion agassizi* in central Eleuthera. Our hypothesis of *Cerion*'s general pattern required that two predictions be affirmed (or else we were in trouble): this population must disappear by hybridization with mottled shells toward bank-interior coasts and with ribby snails toward the bank-edge. We hiked west toward the bank-interior and easily found hybrids right on the verge of the airport road. We then moved east toward the bank-edge along a disused road with vegetation rising to five feet in the center between the tire paths. We should have found our hybrids, but we did not. The *Cerion agassizi* simply stopped about two hundred yards north of our first ribby *Cerion*. Then we realized that a pond lay just to our east and that ribby forms, with their coastal preferences, might not favor the western side of the pond. We forded the pond and found a classic hybrid zone between *Cerion agassizi* and ribby *Cerions*. (Ribby *Cerion* had just managed to round the south end of the pond, but had not moved sufficiently north along the west side to establish contact with *C. agassizi* populations.) I wanted to shout for joy. Then I thought, "But who can I tell; who cares?" And I answered myself, "I don't have to tell anyone. We have just seen and understood something that no one has ever seen and understood before. What more does a man need?"

An eminent colleague, a fine theoretician who has paid his dues in the field, once said to me, only partly facetiously, that fieldwork is one hell of a way to get information. All that time, effort, and money, often for comparatively little when measured against the hours invested. True enough, especially when I count the hours spent drinking Cuban coffee, the one pleasure of my least favorite place, Miami airport. But all the frustration and dull, repetitive effort vanish to insignificance before the unalloyed joy of finding something new—and this pleasure can be savored nearly every day if one loves the little things as well. To say, "We have discovered it; we understand it; we have made some sense and order of nature's confusion." Can any reward be greater?

12 | Human Equality Is a Contingent Fact of History

History's most famous airplane, Lindbergh's *Spirit of St. Louis,* hangs from the ceiling of Washington's Air and Space Museum, imperceptible in its majesty to certain visitors. Several years ago, a delegation of blind men and women met with the museum's director to discuss problems of limited access. Should we build, he asked, an accurate scale model of Lindbergh's plane, freely available for touch and examination? Would such a replica solve the problem? The delegation reflected together and gave an answer that moved me deeply for its striking recognition of universal needs. Yes, they said, such a model would be acceptable, but only on one condition—that it be placed directly beneath the invisible original.

Authenticity exerts a strange fascination over us; our world does contain sacred objects and places. Their impact cannot be simply aesthetic, for an ersatz absolutely indistinguishable from the real McCoy evokes no comparable awe. The jolt is direct and emotional—as powerful a feeling as anything I know. Yet the impetus is purely intellectual—a visceral disproof of romantic nonsense that abstract knowledge cannot engender deep emotion.

Last night, I watched the sun set over the South African savanna—the original location and habitat of our australopithecine ancestors. The air became chill; sounds of the night began, the incessant repetition of toad and insect,

185

laced with an occasional and startling mammalian growl; the Southern Cross appeared in the sky, with Jupiter, Mars, and Saturn ranged in a line above the arms of Scorpio. I sensed the awe, fear, and mystery of the night. I am tempted to say (describing emotions, not making any inferences about realities, higher or lower) that I felt close to the origin of religion as a historical phenomenon of the human psyche. I also felt kinship in that moment with our most distant human past—for an *Australopithecus africanus* may once have stood, nearly three million years ago, on the same spot in similar circumstances, juggling (for all I know) that same mixture of awe and fear.

I was then rudely extricated from that sublime, if fleeting, sentiment of unity with all humans past and present. I remembered my immediate location—South Africa, 1984 (during a respite in Kruger Park from a lecture tour on the history of racism). I also understood, in a more direct way than ever before, the particular tragedy of the history of biological views about human races. This history is largely a tale of division—an account of barriers and ranks erected to maintain the power and hegemony of those on top. The greatest irony of all presses upon me: I am a visitor in the nation most committed to myths of inequality—yet the savannas of this land staged an evolutionary story of opposite import.

My visceral perception of brotherhood harmonizes with our best modern biological knowledge. Such union of feeling and fact may be quite rare, for one offers no guide to the other (more romantic twaddle aside). Many people think (or fear) that equality of human races represents a hope of liberal sentimentality probably squashed by the hard realities of history. They are wrong.

This essay can be summarized in a single phrase, a motto if you will: *Human equality is a contingent fact of history.* Equality is not true by definition; it is neither an ethical principle (though equal treatment may be) nor a statement about norms of social action. It just worked out that way. A hundred different and plausible scenarios for human history would have yielded other results (and moral dilemmas of enormous magnitude). They didn't happen.

The history of Western views on race is a tale of denial
—a long series of progressive retreats from initial claims for
strict separation and ranking by intrinsic worth toward an
admission of the trivial differences revealed by our contingent history. In this essay, I shall discuss just two main
stages of retreat for each of two major themes: genealogy,
or separation among races as a function of their geological
age; and geography, or our place of origin. I shall then
summarize the three major arguments from modern biology for the surprisingly small extent of human racial differences.

GENEALOGY, THE FIRST ARGUMENT

Before evolutionary theory redefined the issue irrevocably,
early to mid-nineteenth-century anthropology conducted a
fierce debate between schools of monogeny and polygeny.
Monogenists advocated a common origin for all people in
the primeval couple, Adam and Eve (lower races, they then
argued, had degenerated further from original perfection).
Polygenists held that Adam and Eve were ancestors of white
folks only, and that other—and lower—races had been separately created. Either argument could fuel a social doctrine
of inequality, but polygeny surely held the edge as a compelling justification for slavery and domination at home and
colonialism abroad. "The benevolent mind," wrote Samuel
George Morton (a leading American polygenist) in 1839,
"may regret the inaptitude of the Indian for civilization.
. . . The structure of his mind appears to be different from
that of the white man. . . . They are not only averse to the
restraints of education, but for the most part are incapable
of a continued process of reasoning on abstract subjects."

GENEALOGY, THE SECOND ARGUMENT

Evolutionary theory required a common origin for human
races, but many post-Darwinian anthropologists found a
way to preserve the spirit of polygeny. They argued, in a
minimal retreat from permanent separation, that the division of our lineage into modern races had occurred so long

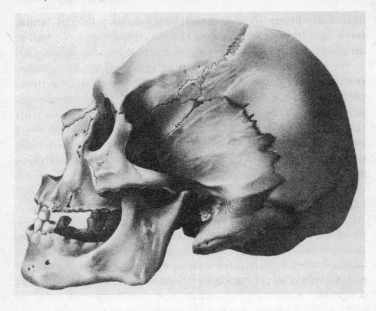

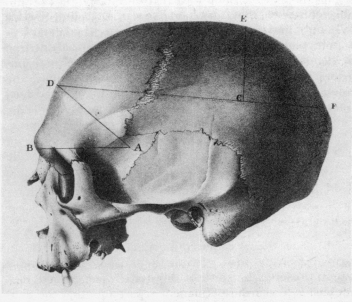

ago that differences, accumulating slowly through time, have now built unbridgeable chasms. Though once alike in an apish dawn, human races are now separate and unequal.

We cannot understand much of the history of late nineteenth- and early twentieth-century anthropology, with its plethora of taxonomic names proposed for nearly every scrap of fossil bone, unless we appreciate its obsession with the identification and ranking of races. For many schemes of classification sought to tag the various fossils as ancestors of modern races and to use their relative age and apishness as a criterion for racial superiority. Piltdown, for example, continued to fool generations of professionals partly because it fit so comfortably with ideas of white superiority. After all, this "ancient" man with a brain as big as ours (the product, we now know, of a hoax constructed with a modern cranium) lived in England—an obvious ancestor for whites —while such apish (and genuine) fossils as *Homo erectus* inhabited Java and China as putative sources for Orientals and other peoples of color.

This theory of ancient separation received its last prominent defense in 1962, when Carleton Coon published his *Origin of Races.* Coon divided humanity into five major races —caucasoids, mongoloids, australoids, and, among African blacks, congoids and capoids. He claimed that these five groups had already become distinct subspecies during the reign of our ancestor, *Homo erectus. H. erectus* then evolved toward *H. sapiens* in five parallel streams, each traversing the

Facing page

(*Top*) One of John Collins's excellent lithographs from S.G. Morton's *Crania Americana,* 1839. Note the subtle "arrangement" chosen (probably unconsciously) to enhance the impression of inferiority in this American Indian skull. The skull is not oriented in the conventional plane, but tilted back, so that the cranium seems lower and less vaulted (therefore smaller and more "primitive"). Compare with next figure of a white male.

(*Bottom*) Skull of a white male (with craniometric measures shown) from Morton's *Crania Americana.* This skull is oriented conventionally (compare with above figure).

same path toward increased consciousness. But whites and yellows, who "occupied the most favorable of the earth's zoological regions," crossed the threshold to *H. sapiens* first, while dark peoples lagged behind and have paid for their sluggishness ever since. Black inferiority, Coon argues, is nobody's fault, just an accident of evolution in less challenging environments:

> Caucasoids and Mongoloids . . . did not rise to their present population levels and positions of cultural dominance by accident. . . . Any other subspecies that had evolved in these regions would probably have been just as successful.

Leading evolutionists throughout the world reacted to Coon's thesis with incredulity. Could modern races really be identified at the level of *H. erectus?* I shall always be grateful to W.E. Le Gros Clark, England's greatest anatomist at the time. I was spending an undergraduate year in England, an absolute nobody in a strange land. Yet he spent an afternoon with me, patiently answering my questions about race and evolution. Asked about Coon's thesis, this splendidly modest man simply replied that he, at least, could not identify a modern race in the bones of an ancient species.

More generally, parallel evolution of such precision in so many lineages seems almost impossible on grounds of mathematical probability alone. Could five separate subspecies undergo such substantial changes and yet remain so similar at the end that all can still interbreed freely, as modern races so plainly do? In the light of these empirical weaknesses and theoretical implausibilities, we must view Coon's thesis more as the last gasp of a dying tradition than a credible synthesis of available evidence.

GENEALOGY, THE MODERN VIEW

Human races are not separate species (the first argument) or ancient divisions within an evolving plexus (the second argument). They are recent, poorly differentiated subpopulations of our modern species, *Homo sapiens,* separated at most by tens or hundreds of thousands of years, and marked by remarkably small genetic differences.

GEOGRAPHY, THE FIRST ARGUMENT

When Raymond Dart found the first australopithecine in South Africa sixty years ago, scientists throughout the world rejected this oldest ancestor, this loveliest of intermediate forms, because it hailed from the wrong place. Darwin, without a shred of fossil evidence but with a good criterion for inference, had correctly surmised that humans evolved in Africa. Our closest living relatives, he argued, are chimps and gorillas—and both species live only in Africa, the probable home, therefore, of our common ancestor as well.

But few scientists accepted Darwin's cogent inference because hope, tradition, and racism conspired to locate our ancestral abode on the plains of central Asia. Notions of Aryan supremacy led anthropologists to assume that the vast "challenging" reaches of Asia, not the soporific tropics of Africa, had prompted our ancestors to abandon an apish past and rise toward the roots of Indo-European culture. The diversity of colored people in the world's tropics could only record the secondary migrations and subsequent degenerations of this original stock. The great Gobi Desert expedition, sponsored by the American Museum of Natural History just a few years before Dart's discovery, was dispatched primarily to find the ancestry of man in Asia. We remember this expedition for success in discovering dinosaurs and their eggs; we forget that its major quest ended in utter failure because Darwin's simple inference was correct.

GEOGRAPHY, THE SECOND ARGUMENT

By the 1950s, further anatomical study and the sheer magnitude of continuing discovery forced the general admission that our roots lay with the australopithecines, and that Africa had been our original home. But the subtle hold of unacknowledged prejudice still conspired (with other, more reasonable bases of uncertainty) to deny Africa its continuing role as the cradle of what really matters to us—the origin of human consciousness. In a stance of intermediate retreat, most scientists now argued that Africa had kindled our origin but not our mental emergence. Human ancestors migrated out, again to mother Asia, and there crossed the threshold to consciousness as *Homo erectus* (or so-called Java and Peking man). We emerged from the apes in Africa; we evolved our intelligence in Asia. Carleton Coon wrote in his 1962 book: "If Africa was the cradle of mankind, it was only an indifferent kindergarten. Europe and Asia were our principal schools."

GEOGRAPHY, THE MODERN VIEW

The tempo of African discovery has accelerated since Coon constructed his metaphor of the educational hierarchy. *Homo erectus* apparently evolved in Africa as well, where fossils dating to nearly two million years have been found, while the Asian sites may be younger than previously imagined. One might, of course, take yet another step in retreat and argue that *H. sapiens,* at least, evolved later from an Asian stock of *H. erectus.* But the migration of *H. erectus* into Europe and Asia does not guarantee (or even suggest) any further branching from these mobile lineages. For *H. erectus* continued to live in Africa as well. Evidence is not yet conclusive, but the latest hints may be pointing toward an African origin for *H. sapiens* as well. Ironically then (with respect to previous expectations), every human species may have evolved first in Africa and only then—for the two latest species of *Homo*—spread elsewhere.

I have, so far, only presented the negative evidence for my thesis that human equality is a contingent fact of history. I have argued that the old bases for inequality have evaporated. I must now summarize the positive arguments (primarily three in number) and, equally important, explain how easily history might have happened in other ways.

THE POSITIVE (AND FORMAL, OR TAXONOMIC) ARGUMENT FROM RACIAL DEFINITION

We recognize only one formal category for divisions within species—the subspecies. Races, if formally defined, are therefore subspecies. Subspecies are populations inhabiting a definite geographic subsection of a species' range and sufficiently distinct in any set of traits for taxonomic recognition. Subspecies differ from all other levels of the taxonomic hierarchy in two crucial ways. First, they are categories of convenience only and need never be designated. Each organism must belong to a species, a genus, a family, and to all higher levels of the hierarchy; but a species need not be formally divided. Subspecies represent a taxonomist's personal decision about the best way to report geographic variation. Second, the subspecies of any species cannot be distinct and discrete. Since all belong to a single species, their members can, by definition, interbreed. Modern quantitative methods have permitted taxonomists to describe geographic variation more precisely in numerical terms; we need no longer construct names to describe differences that are, by definition, fleeting and changeable. Therefore, the practice of naming subspecies has largely fallen into disfavor, and few taxonomists use the category any more. Human variation exists; the formal designation of races is passé.

Some species are divided into tolerably distinct geographic races. Consider, for example, an immobile species separated on drifting continental blocks. Since these subpopulations never meet, they may evolve substantial differences. We might still choose to name subspecies for such discrete geographic variants. But humans move about and maintain the most notorious habits of extensive interbreeding. We are not well enough divided into distinct geographic groups, and the naming of human subspecies makes little sense.

Our variation displays all the difficulties that make taxonomists shudder (or delight in complexity) and avoid the naming of subspecies. Consider just three points. First, discordance of characters. We might make a reasonable division on skin color, only to discover that blood groups imply different alliances. When so many good characters exhibit such discordant patterns of variation, no valid criterion can be established for unambiguous definition of subspecies. Second, fluidity and gradations. We interbreed wherever we move, breaking down barriers and creating new groups. Shall the Cape Colored, a vigorous people more than two million strong and the offspring of unions between Africans and white settlers (the ancestors, ironically, of the authors of apartheid and its antimiscegenation laws), be designated a new subspecies or simply the living disproof that white and black are very distinct? Third, convergences. Similar characters evolve independently again and again; they confound any attempt to base subspecies on definite traits. Most indigenous tropical people, for example, have evolved dark skin.

The arguments against naming human races are strong, but our variation still exists and could, conceivably, still serve as a basis for invidious comparisons. Therefore, we must add the second and third arguments as well.

THE POSITIVE ARGUMENT FROM RECENCY OF DIVISION

As I argued in the first part of this essay (and need only state in repetition now), the division of humans into modern "racial" groups happened yesterday, in geological terms. This differentiation does not predate the origin of our own species, *Homo sapiens,* and probably occurred during the last few tens (or at most hundreds) of thousands of years.

THE POSITIVE ARGUMENT FROM GENETIC SEPARATION

Mendel's work was rediscovered in 1900 and the science of genetics spans our entire century. Yet, until twenty years ago, a fundamental question in evolutionary genetics could not be answered for a curious reason. We were not able to calculate the average amount of genetic difference between organisms because we had devised no method for taking a random sample of genes. In the classical Mendelian analysis of pedigrees, a gene cannot be identified until it varies among individuals. For example, if absolutely every *Drosophila* in the world had red eyes, we would rightly suspect that some genetic information coded this universal feature, but we would not be able to identify a gene for red eyes by analyzing pedigrees, because all flies would look the same. But as soon as we find a few white-eyed flies, we can mate white with red, trace pedigrees through generations of offspring, and make proper inferences about the genetic basis of eye color.

To measure the average genetic differences among races, we must be able to sample genes at random—and this unbiased selection can't be done if we can only identify variable genes. Ninety percent of human genes might be held in common by all people, and an analysis confined to varying genes would grossly overestimate the total difference.

In the late 1960s, several geneticists harnessed the common laboratory technique of electrophoresis to solve this old dilemma. Genes code for proteins, and varying proteins may behave differently when subjected in solution to an electric field. Any protein could be sampled, independent of prior knowledge about whether it varied or not. (Electrophoresis can only give us a minimal estimate because some varying proteins may exhibit the same electrical mobility but be different in other ways.) Thus, with electrophoresis we could finally ask the key question: How much genetic difference exists among human races?

The answer, surprising for many people, soon emerged without ambiguity: damned little. Intense studies for more than a decade have detected not a single "race gene"—that is, a gene present in all members of one group and none of another. Frequencies vary, often considerably, among groups, but all human races are much of a muchness. We can measure so much variation among individuals *within* any race that we encounter very little new variation by adding another race to the sample. In other words, the great preponderance of human variation occurs within groups, not in the differences between them. My colleague Richard Lewontin (see bibliography), who did much of the original electrophoretic work on human variation, puts it dramatically: If, God forbid, the holocaust occurs "and only the Xhosa people of the southern tip of Africa survived, the human species would still retain 80 percent of its genetic variation."

As long as most scientists accepted the ancient division of races, they expected important genetic differences. But the recent origin of races (second positive argument) affirms the minor genetic differences now measured. Human groups do vary strikingly in a few highly visible characters (skin color, hair form)—and these external differences may fool us into thinking that overall divergence must be great. But we now know that our usual metaphor of superficiality —skin deep—is literally accurate.

In thus completing my précis, I trust that one essential point will not be misconstrued: I am, emphatically, not talking about ethical precepts but about information in our best current assessment. It would be poor logic and worse strategy to hinge a moral or political argument for equal treatment or equal opportunity upon any factual statement about human biology. For if our empirical conclusions need revision—and all facts are tentative in science—then we might be forced to justify prejudice and apartheid (directed, perhaps, against ourselves, since who knows who would turn up on the bottom). I am no ethical philosopher, but I can only view equality of opportunity as inalienable, universal, and unrelated to the biological status of individuals. Our races may vary little in average characters, but our individuals differ greatly—and I cannot imagine a decent world that does not treat the most profoundly retarded person as a full human being in all respects, despite his evident and pervasive limitations.

I am, instead, making a smaller point, but one that tickles my fancy because most people find it surprising. The conclusion is evident, once articulated, but we rarely pose the issue in a manner that lets such a statement emerge. I have called equality among races a *contingent* fact. So far I have only argued for the fact; what about the contingency? In other words, how might history have been different? Most of us can grasp and accept the equality; few have considered the easy plausibility of alternatives that didn't happen.

My creationist incubi, in one of their most deliciously ridiculous arguments, often imagine that they can sweep evolution away in the following unanswerable riposte: "Awright," they exclaim, "you say that humans evolved from apes, right?" "Right," I reply. "Awright, if humans evolved from apes, why are apes still around? Answer that one!" If evolution proceeded by this caricature—like a ladder of progress, each rung disappearing as it transforms bodily to the next stage—then I suppose this argument would merit attention. But evolution is a bush, and ancestral groups usually survive after their descendants branch off. Apes come in many shapes and sizes; only one line led to modern humans.

Most of us know about bushes, but we rarely consider the implications. We know that australopithecines were our ancestors and that their bush included several species. But we view them as forebears, and subtly assume that since we are here, they must be gone. It is so indeed, but it ain't *necessarily* so. One population of one line of australopithecines became *Homo habilis;* several others survived. One species, *Australopithecus robustus,* died less than a million years ago and lived in Africa as a contemporary of *H. erectus* for a million years. We do not know why *A. robustus* disappeared. It might well have survived and presented us today with all the ethical dilemmas of a human species truly and markedly inferior in intelligence (with its cranial capacity only one-third our own). Would we have built zoos, established reserves, promoted slavery, committed genocide, or perhaps even practiced kindness? Human equality is a contingent fact of history.

Other plausible scenarios might also have produced marked inequality. *Homo sapiens* is a young species, its division into races even more recent. This historical context has not provided enough time for the evolution of substantial differences. But many species are millions of years old, and their geographic divisions may be marked and deep. *H. sapiens* might have evolved along such a scale of time and produced races of great age and large accumulated differences—but we didn't. Human equality is a contingent fact of history.

A few well-placed mottoes might serve as excellent antidotes against deeply ingrained habits of Western thought that so constrain us because we do not recognize their influence—so long as these mottoes become epitomes of real understanding, not the vulgar distortions that promote "all is relative" as a précis of Einstein.

I have three favorite mottoes, short in statement but long in implication. The first, the epitome of punctuated equilibrium, reminds us that gradual change is not the only reality in evolution: other things count as well; "stasis is data." The second confutes the bias of progress and affirms that evolution is not an inevitable sequence of ascent: "mammals evolved at the same time as dinosaurs." The third is the theme of this essay, a fundamental statement about human variation. Say it five times before breakfast tomorrow; more important, understand it as the center of a network of implication: "Human equality is a contingent fact of history."

13 | The Rule of Five

THE HUMAN MIND DELIGHTS in finding pattern
—so much so that we often mistake coincidence or forced
analogy for profound meaning. No other habit of thought
lies so deeply within the soul of a small creature trying to
make sense of a complex world not constructed for it.

Into this Universe, and why not knowing
Nor whence, like water willy-nilly flowing,

as the *Rubáiyát* says. No other error of reason stands so
doggedly in the way of any forthright attempt to understand
some of the world's most essential aspects—the tortuous
paths of history, the unpredictability of complex systems,
and the lack of causal connection among events superficially
similar.

Numerical coincidence is a common path to intellectual
perdition in our quest for meaning. We delight in catalogs
of disparate items united by the same number, and often
feel in our gut that some unity must underlie it all. Our
ancestors pondered the mystique of seven—the number of
planets (sun, moon, and five visible planets, all circling the
earth in Ptolemy's system), the deadly sins, the seals of
Revelations. Five has also been favored, not only for fingers
and toes but also for the acts of a proper play according to
Horace, the smooth stones that David selected to slay Goli-
ath, the few loaves that Christ used to feed the multitude,

the number of Mrs. Bixby's sons (all of whom, apparently, did not die gloriously on the field of battle, Mr. Lincoln notwithstanding). The owl and pussycat went to sea with all their worldly goods wrapped in a five-pound note (a very large bill—in physical size as well as monetary value—in those Victorian days). What this country needs, and will never have again, is a good five-cent cigar.

In this essay, I shall discuss two taxonomic systems (theories for the classification of organisms) popular in the decades just before Darwin published the *Origin of Species*. Both assumed reasons other than evolution for the ordering of organisms; both proposed a scheme based on the number five for placing organisms into a hierarchy of groups and subgroups. Both argued that such a simple numerical regularity must record an intrinsic pattern in nature, not a false order imposed by human hope upon a more complex reality. I shall describe these systems and then discuss how evolutionary theory undermined their rationale and permanently changed the science of taxonomy by making such simple numerical schemes inconsistent with a new view of nature. This important change in scientific thought embodies a general message about the character of history and the kinds of order that a world built by history, and not by preordained plan, can (and cannot) express.

Louis Agassiz wrote of his teacher, the German embryologist Lorenz Oken:

> A master in the art of teaching, he exercised an almost irresistible influence over his students. Constructing the universe out of his own brain . . . classifying the animals as if by magic, in accordance with an analogy based on the dismembered body of man, it seemed to us who listened that the slow laborious process of accumulating precise detailed knowledge could only be the work of drones, while a generous, commanding spirit might build the world out of its own powerful imagination.

Oken was a fine descriptive anatomist; his treatises on the embryology of the pig and dog, written in 1806, are classics of meticulous care. But Oken was also a leader in the popular early nineteenth-century school of *Naturphilosophie*—an intellectual movement based on the romantic vision that simple laws of dynamic motion ruled nature and that great intellects might apprehend these laws by a kind of creative intuition. Oken's major contribution to this movement, his *Lehrbuch der Naturphilosophie* (1809–1811), is a list, close to 4,000 items long, taking all knowledge for its province and full of oracular pronouncements about nearly everything, from why the earth is a crystal (with mountain ranges as its edges) to why *Kriegskunst* (the art of war) is the noblest of human endeavors.

Oken, though widely respected in his day (even by his intellectual enemies), has suffered the fate of modern citation largely for laughs about the bad old past versus the bright present. Admittedly, his oracular style of pronouncement invites ridicule by modern standards. What else can one make of Oken's paean to zero: "The whole of mathematics emerges out of zero, so must everything . . . have emerged from the eternal or nothing of nature. . . . There is no other science than that which treats of nothing." Or his claim that all "lower" animals are merely incomplete humans: "The animal kingdom is only a dismemberment of the highest animal, that is, of Man."

When divorced from Oken's own context, these statements lose all meaning, and we can only laugh at their disembodied style. When properly situated, they at least make sense (though we may judge them incorrect today), and we may attribute Oken's peculiar style to differences in taste and custom, not to stupidity or irrelevance.

The context for most of his peculiar pronouncements— the primacy of zero, animals as aborted humans, taxonomy by fives—lies in the primary doctrine of *Naturphilosophie:* the idea of a single, progressive developmental tendency in nature. All natural processes move upward in one direction, starting from primal nothingness (Oken's zero) and advancing toward human complexity and beyond. (Oken's view is not evolutionary since each new state begins again in the primal zero and moves one step beyond its predecessor. A higher form does not evolve by direct genealogical descent from a less developed ancestor, as an evolutionary theory would require.) Since all animals can be arranged in a single series of ascending complexity, with humans on top, lower creatures are incomplete humans. (Oken defined each new step in complexity as the *addition* of an organ; thus, creatures below us on the ladder of progress contain fewer organs and are incomplete.)

The excitement of new theories lies in their power to change contexts, to render irrelevant what once seemed sensible. If we laugh at the past because we judge it anachronistically in the light of present theories, how can we understand these changes of context? And how can we retain proper humility toward our own favored theories and the probability of their own future lapse into insignificance? Honest intellectual passions always merit respect.

Evolutionary theory was the greatest context changer in the history of biology. Theodosius Dobzhansky wrote in a famous statement that nothing in biology makes sense except in the light of evolution. But Oken's world made sense under a different set of beliefs about how nature worked. Dobzhansky meant, of course, that once we recognize evolution as the basis of organic history, all biology must be reformulated. But if we wish to understand why evolution was such a watershed in the history of ideas, we must comprehend the contexts that it replaced, not view them as imperfect harbingers of evolution. They were different, subtle, brilliant (and wrong), not stupid. We must study such theories as Oken's classification by fives, and we must learn why evolution destroyed their rationale if we wish to grasp the sweep and power of evolution itself.

Oken's taxonomy by fives attempts to reconcile two principles, both dear to *Naturphilosophie,* but superficially in contradiction—first, that animals represent a single series of increasing complexity defined by the successive addition of organs; second, that meaningful analogies permeate nature and that each segment of taxonomy mimics or mirrors all the others (the order of mammals, for example, must repeat in miniature the same scheme that arranges all of nature). But how can nature contain, simultaneously, a single ascending series and a set of repeating cycles?

Let us examine the two claims separately. Consider first Oken's epitome of his belief that all animals form a single series marked by the addition of organs, aphorisms 3067–3072 of his *Lehrbuch:*

> The animal kingdom is only one animal. . . .
>
> The animal kingdom is only a dismemberment of the highest animal, that is, of Man.
>
> Animals become nobler in rank, the greater the number of organs that are collectively liberated or severed from the Grand Animal, and that enter into combination. An animal that, for example, lived only as intestine, would be, doubtless, inferior to one that with the intestine were to combine a skin. . . .
>
> Animals are gradually perfected . . . by adding organ unto organ. . . .
>
> Each animal ranks therefore above the other; two of them never stand on an equal plane or level.
>
> Animals are distinguished . . . by the number of their different organs.

But such a simple linear order could not satisfy the soul of a man who believed that every nuance of nature had deep meaning in its union with all other parts. Oken could not leave the amoeba in its pond or the crab on the seashore, for all creatures must be elements of a complex and interconnected harmony, not merely the lower rungs of a ladder. Thus, Oken developed a scheme for crossties; he would classify nature as a web of meaning, not just a line of progress.

Oken felt that he had broken the code of nature's numerical order in recognizing pervasive cycles of five based upon the organs of sense in their own ascending sequence: feeling, taste, smell, hearing, and sight. Driven by his romantic vision of living matter yearning for perfection along simple

paths bursting with meaning, Oken found ascending circles
of five everywhere, from the grandest scale of all animals to
the smallest of human races.

He arranged the entire animal kingdom in a rising cycle
of five, reflecting the successive addition (or perfection) of
sensory organs. "The animal classes," he wrote, "are virtually nothing else than a representation of the sense organs.
. . . They must be arranged in accordance with them."
Invertebrates, fishes, reptiles, birds, and mammals, or feeling, taste, smell, hearing, and sight. I shall not burden this
essay with Oken's forced and specious arguments for these
fanciful correspondences. Recalcitrant, complex nature
behaves very badly whenever we try to force such simple
schemes upon her (consider, for example, the difficulty of
identifying mammals with sight, when the lower class of
birds contains species with vision more acute than any
mammal's). I shall simply cite Oken's rationale.

Strictly speaking, there are only 5 animal classes: Der-
matozoa, or the Invertebrata; Glossozoa, or the
Fishes, as being those animals in whom a true tongue
makes for the first time its appearance; Rhinozoa, or
the Reptiles, wherein the nose opens for the first time
into the mouth and inhales air; Otozoa, or the Birds,
in which the ear for the first time opens externally;
Ophthalmozoa, or the Thricozoa [mammals], in
whom all the organs of sense are present and complete, the eyes being moveable and covered with two
palpebrae or lids.

As for the large, so for the small. Oken even managed to
portray the conventional racist ordering of human groups
by his sensory analogy, although he didn't even attempt a
rationale for his choices:

1. The skin-man is the black, African
2. The tongue-man is the brown, Australian-Malayan
3. The nose-man is the red, American

4. The ear-man is the yellow, Asiatic-Mongolian
5. The eye-man is the white, European.

But how can nature move in cycles regulated by organs of sense and, at the same time, along a single path of progress governed by the addition of organs? We need an image, an analogy, and a chart.

Image: The object that moves up the path of progress is not a striding creature, but a rolling circle with five spokes marked feeling, taste, smell, hearing, and sight. Each time a spoke touches the ground, it deposits a creature representing its level of sensory advance along the path of progress. When the highest spoke of sight finally reaches the ground, a new and smaller wheel starts rolling again, depositing creatures along the same sensory sequence.

Analogy: Several theories of history in Western thought manage to unite ideas of continuous progress with cyclical repetitions. In the sixteenth-century glass of King's College Chapel, Cambridge, a powerful figure of Jonah, belched forth from the whale's belly, overlies an image of Christ rising from the tomb—for both men came to life again on the third day *in extremis.* Christian history moves inexorably forward, but the New Testament replays the Old, and God's meaning lies revealed in the repetition.

Chart: The following chart shows four cycles of five-part sensory wheels: all animals, all mammals, the highest group of mammals, and the highest species of the highest group. For Oken, these identifications with sense organs and specification of five-part wheels at all scales throughout nature did not represent an artificial system constructed to aid memory or facilitate recall, but a discovery of nature's underlying reality. He expected practical results from his correspondences. He tried, for example, to arrange the mineral and vegetable world in five-part wheels as well. Since our medicines are made of chemicals and plants, the correct correspondences will specify proper treatments. We might cure Africans with the plants of feeling, Caucasians, with plants of sight.

Oken's Path of Progress by Wheels of Five

ALL ANIMALS		MAMMALS		ADVANCED MAMMALS		HUMANS	
Feeling	Invertebrates						
Taste	Fishes						
Smell	Reptiles						
Hearing	Birds						
Sight	Mammals	Feeling	rodents				
		Taste	sloths & marsupials				
		Smell	bats & insectivores				
		Hearing	whales & hoofed mammals				
		Sight	carnivores & primates	Feeling	cats & dogs		
				Taste	seals		
				Smell	bears		
				Hearing	apes		
				Sight	humans	Feeling	Africans
						Taste	Australians
						Smell	Americans
						Hearing	Orientals
						Sight	Caucasians

REPRINTED FROM NATURAL HISTORY.

If once the genera of Minerals, Plants and Animals come to stand correctly opposite each other, a great advantage will accrue therefrom to the science of Materia Medica; for corresponding genera will act specifically upon each other.

I admire the sweep and coherence of Oken's vision, but I'll descend to the realm of sound and be a monkey's uncle if it says anything much about nature.

As Oken constructed his ascending wheels of five in Germany during the decades before Darwin, another taxonomic theory, the quinary system, led many English naturalists to arrange all organisms into different circles of five. The quinary system invites comparison with Oken's scheme because it also built circles of five at different scales and sought correspondences between organisms at the same position on different circles. It also attempted to resolve the apparent contradiction between linear progress and circular repetition.

The quinary system rests upon a separation between two

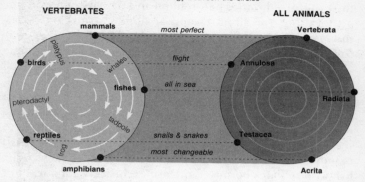

Swainson's quinary circles for vertebrates & all animals, showing ties of affinity for vertebrates & ties of analogy between the circles

REPRINTED FROM NATURAL HISTORY.

kinds of similarity: affinity and analogy. Ties of affinity unite forms on the same circle; analogies specify correspondence between circles. For example, William Swainson, a leading British quinarian, justified the following circle of vertebrates in 1835. We recognize fish, amphibians, reptiles, birds, and mammals as five groups of common anatomical design. But how can they represent both an ascending pathway and a closed circle of five? Swainson argues that we must unite each pair by an intermediate form showing ties of affinity—fish to amphibian by the tadpole, amphibian to reptile by the adult frog, reptile to bird by the flying pterodactyl, bird to mammal by the duck-billed platypus, and mammal back to fish by the largest agent of natural transport, whales. Since whales connect the highest mammals to the lowest fish, the path of progress curves back on itself and forms a circle. "Nature herself," Swainson proclaimed, "describes the mighty circle and pronounces it complete."

The circle of vertebrates can then be united with other circles at both smaller and larger scales by ties of analogy linking groups in similar positions. (I must confess that Swainson's arguments seem as forced as Oken's. The quinarians never presented rigorous criteria for why certain relationships should be called affinity and others analogy.

One gets the uncomfortable feeling that they constructed their preferred circles beforehand and then invented ad hoc justifications for affinities and analogies so ordained—although the method supposedly worked the other way round, building circles and correspondences from raw data of affinity and analogy.) For example, Swainson arranged all animals into a circle of Radiata (echinoderms and their kin), Acrita (protozoans and other "simple" creatures), Testacea (mollusks), Annulosa (segmented worms, insects, and crustaceans), and Vertebrata. The supposed ties of analogy to the vertebrate circle seem a bit contrived, to say the least: mammals with vertebrates as the most perfect of each circle; fish with radiates because both are exclusively aquatic, "not one species in either group having yet been discovered upon the land"; Amphibia with Acrita because both (get this) "however dissimilar in other respects are remarkable for changing their shapes more than any other of the aberrant types in either circle"; reptiles with mollusks because both serpents and snails lack feet and crawl on their bellies; and birds with Annulosa because insects fly too.

I was disappointed to discover that the article on Swainson in the historians' bible, the *Dictionary of Scientific Biography* (called the *DSB* by all pros), follows the old tradition, criticized earlier in this essay, of dismissing superseded systems as pathetically foolish in the light of modern knowledge:

> His indefatigable pursuit of natural history and conscientious labor on its behalf deserve to be remembered as a set-off against the injury he unwittingly caused by his adherence to the absurd quinary system. . . . This extraordinary theory was pertinaciously held by Swainson throughout his zoological career and it certainly impaired much of his work.

Oken and Swainson were severely and legitimately criticized in their own terms. (I have tried to raise some of these arguments by exposing the fanciful criteria used to establish circles of five and to draw analogies between them.) But they were not fools or madmen, and their systems were not absurd. Oken and Swainson ranked highly among the best natural historians of Europe, and their numerical systems of taxonomy were popular and serious contenders among contemporary schemes for ordering nature.

Rigid numerical systems only become absurd in the later light of evolution, for their respectability hinges upon *theories* favored for the causes of nature's order. If God placed species on earth (as Swainson believed), then he might have acted with a numerical precision displaying the rigor and harmony of his thoughts. If simple laws, rather than accidents of history, establish the sequence of organisms (as Oken held), then numerical order might arise among animals just as the periodic table regulates chemical elements. Numerology in taxonomy may be dismissed as absurd mysticism today, but in Oken and Swainson's time, this approach embodied a reasonable result of defensible theories about the causes of nature's order. Swainson put it right on the line when he inferred from quinary order both God's existence and his special concern for us:

> When we discover evident indications of a definite plan, upon which all these modifications have been regulated by a few simple and universal laws, our wonder is as much excited at the inconceivable wisdom and goodness of the SUPREME by whom these myriads of beings have been created and are now preserved, as at the mental blindness and perverted understanding of those philosophers, falsely so called, who would persuade us, that even Man, the last and best of created things, is too insignificant for the special care of Omnipotence.

Darwin destroyed the rule of five forever because he removed its rationale by reconstructing nature. His agent of destruction was not evolution itself. I can imagine evolutionary theories (indeed some have been proposed) so committed to foreordination by simple laws or directing intelligences that numerical order might still emerge from rigidly predictable process. Darwin's exterminating angel was, simply, history. Evolution does not unroll according to simple laws specifying necessary results. It follows the vagaries of history. Its pathways are twisted and churned by changing environments, from minor shifts in temperature and precipitation to the rise of mountain chains, the growth of glaciers, the drift of continents, and even (probably) the impact of comets or asteroids. Evolution cannot achieve engineering perfection because it must work with inherited parts available from previous histories in different contexts: the panda's "thumb" is a clumsy, detached wrist bone, pressed into service because the true first digit became committed to other functions during ancestral life as a conventional carnivore; we suffer the pain of aching backs and the annoyances of hernias because large four-footed creatures of our lineage were not really made to walk on their toes—four legs good, two legs not so good.

How could animals evolve along the tortuous pathways of history and arrange themselves neatly into circles of five? Numerical precision cannot regulate taxonomy because life unfolds in time. Evolution records a complex, irrevocable history; its pathways were not preordained by simple rules or commanding intelligences.

But life regulated by history still has order—firm, ineluctable, definite, testable pattern. Its order is the topology of its proper metaphor—the tree of life. Its order is genealogy, connection by branching and descent. Swainson *described* the biological world correctly before he went too far:

> Had the order of nature been so irregular that we had found she created some birds with four feet, others with two, and some with none; or that, like the fabulous griffin, there were creatures half quadruped, half bird; or, if insects had been found with the feet of quadrupeds, and the toes of birds; in short, had such compounds in the animal world existed, the foundations of natural history, as a science, could never have been laid.

Darwin then found the *reason* for order and changed our world forever:

> Something more is included in our classification than mere resemblance. I believe that something more is . . . propinquity of descent,—the only known cause of the similarity of organic beings.

4 | Trends and Their Meaning

14 | Losing the Edge

I WISH TO PROPOSE a new kind of explanation for the oldest chestnut of the hot stove league—the most widely discussed trend in the history of baseball statistics: the extinction of the .400 hitter. Baseball aficionados wallow in statistics, a sensible obsession that outsiders grasp with difficulty and ridicule often. The reasons are not hard to fathom. In baseball, each essential action is a contest between two individuals—batter and pitcher, or batter and fielder—thus creating an arena of truly individual achievement within a team sport.

The abstraction of personal achievement in other team sports makes comparatively little sense. Goals scored in basketball or yards gained in football depend on the indissoluble intricacy of team play; a home run is you against him. Moreover, baseball has been played under a set of rules and conditions sufficiently constant during our century to make comparisons meaningful, yet sufficiently different in detail to provide endless grist for debate (the "dead ball" of 1900–1920 versus the "lively ball" of later years, the introduction of night games and relief pitchers, the invention of the slider, the changing and irregular sizes of ball parks, nature's own versus Astroturf).

No subject has inspired more argument than the decline and disappearance of the .400 hitter—or, more generally, the drop in league-leading batting averages during our century. Since we wallow in nostalgia and have a lugubrious tendency to compare the present unfavorably with a past "golden era," this trend acquires all the more fascination because it carries moral implications linked metaphorically with junk foods, nuclear bombs, and eroding environments as signs of the current decline and impending fall of Western civilization.

Between 1901 and 1930, league-leading averages of .400 or better were common enough (nine out of thirty years) and achieved by several players (Lajoie, Cobb, Jackson, Sisler, Heilmann, Hornsby, and Terry), and averages over .380 scarcely merited extended commentary. Yet the bounty dried up abruptly thereafter. In 1930 Bill Terry hit .401 to become the last .400 hitter in the National League; and Ted Williams's .406 in 1941 marked the last pinnacle for the American League. Since Williams, the greatest hitter I ever saw, attained this goal in the year I was born (and I am, alas, no spring chicken), only three men have hit higher than .380 in a single season: Williams again in 1957 (.388, at age thirty-nine, with my vote for the greatest batting accomplishment of our era), Rod Carew (.388 in 1977), and George Brett (.390 in 1980). Where have all the hitters gone?

League averages for our century

	American League	National League
1901–1910	.251	.253
1911–1920	.259	.257
1921–1930	.286	.288
1931–1940	.279	.272
1941–1950	.260	.260
1951–1960	.257	.260
1961–1970	.245	.253
1971–1980	.258	.256

Two rather different kinds of explanation have been traditionally offered. The first, naive and moral, simply acknowledges with a sigh that there were giants on the earth in those days. Something in us needs to castigate the present in the light of an unrealistically rosy past. In researching the history of misconduct, for example, I discovered that every generation (at least since the mid-nineteenth century) has imagined itself engulfed in a crime wave. Each age has also witnessed a shocking decline in sportsmanship. Similarly, senior citizens of the hot stove league, and younger fans as well (for nostalgia may have its greatest emotional impact on those too young to know a past reality directly), tend to argue that the .400 hitters of old simply cared more and tried harder. Well, Ty Cobb may have been a paragon of intensity and a bastard to boot, and Pete Rose may be a gentleman by comparison, but today's play is anything but lackadaisical. Say what you will; monetary rewards in the millions do inspire single-minded effort.

The second kind of explanation views people as much of a muchness over time and attributes the downward trend in league-leading batting to changes in the game and its styles of play. Most often cited are improvements in pitching and fielding, and more grueling schedules that shave off the edge of excellence. J.L. Reichler, for example, one of baseball's premier record keepers, argues (see bibliography):

> The odds are heavily against another .400 hitter because of the tremendous improvement in relief pitching and fielding. Today's players face the additional handicaps of a longer schedule, which wears down even the strongest players, and more night games, in which the ball is harder to see.

I do not dispute Reichler's factors, but I believe that he offers an incomplete explanation, expressed from an inadequate perspective.

Another proposal in this second category invokes the numerology of baseball. Every statistics maven knows that, following the introduction of the lively ball in the early 1920s (and Babe Ruth's mayhem upon it), batting averages soared in general and remained high for twenty years. As the accompanying chart shows, league averages for all players rose into the .280s in both leagues during the 1920s and remained in the .270s during the 1930s, but never topped

.260 in any other decade of our century. Naturally, if league averages rose so substantially, we should not be surprised that the best hitters also improved their scores. The great age of .400 hitting in the National League did occur during the 1920s (another major episode of high averages occurred in the pre-modern era, during the 1890s, when the decadal average rose to .280—it had been .259 for the 1870s and .254 for the 1880s).

But this simple factor cannot explain the extinction of .400 hitting either. No one hit .400 in either league during 1931–1940, even though league averages stood twenty points above their values for the first two decades of our century, when fancy hitting remained in vogue. A comparison of these first two decades with recent times underscores both the problem and the failure of resolutions usually proposed—for high hitting in general (and .400 hitting in particular) flourished from 1900 to 1920, but league averages back then did not differ from those for recent decades, while high hitting has gone the way of bird's teeth.

Consider, for example, the American League during 1911–1920 (league average of .259) and 1951–1960 (league average of .257). Between 1911 and 1920, averages above .400 were recorded during three years, and the leading average dipped below .380 only twice (Cobb's .368 and .369 in 1914 and 1915). This pattern of high averages was not just Ty Cobb's personal show. In 1912 Cobb hit .410, while the ill-fated Shoeless Joe Jackson reached .395, Tris Speaker .383, thirty-seven-year-old Nap Lajoie .368, and Eddie Collins .348. By comparison, during 1951–1960, only three leading averages exceeded Eddie Collins's fifth-place .348 (Mantle's .353 in 1956, Kuenn's .353 in 1959, and Williams's .388, already discussed, in 1957). The 1950s, by the way, was not a decade of slouches, what with the likes of Mantle, Williams, Minoso, and Kaline. Thus, a general decline in league-leading averages throughout the century cannot be explained by an inflation of general averages during two middle decades. We are left with a puzzle. As with most persistent puzzles, we probably need new *kind* of explanation, not merely a recycling and refinement of old arguments.

I am a paleontologist by trade. We students of life's history spend most of our time worrying about long-term trends. Has life become more complex through time? Do more species of animals live now than 200 million years ago? Several years ago, it occurred to me that we suffer from a subtle but powerful bias in our approach to explaining trends. Extremes fascinate us (the biggest, the smallest, the oldest), and we tend to concentrate on them alone, divorced from the systems that include them as unusual values. In explaining extremes, we abstract them from larger systems and assume that their trends arise for self-generated reasons: if the biggest become bigger through time, some powerful advantage must accompany increasing size.

But if we consider extremes as the limiting values of larger systems, a very different kind of explanation often applies. If the *amount of variation* within a system changes (for whatever reason), then extreme values may increase (if total variation grows) or decrease (if total variation declines) without any special reason rooted in the intrinsic character or meaning of the extreme values themselves. In other words, *trends in extremes* may result from systematic changes in *amounts of variation.* Reasons for changes in variation are often rather different from proposed (and often spurious) reasons for changes in extremes considered as independent from their systems.

Let me illustrate this unfamiliar concept with two examples from my own profession—one for increasing, the other for decreasing extreme values. First, an example of increasing extreme values properly interpreted as an expansion of variation: The largest mammalian brain sizes have increased steadily through time (the brainiest have gotten brainier). Many people infer from this fact that inexorable trends to increasing brain size affect most or all mammalian lineages. Not so. Within many groups of mammals, the most common brain size has not changed at all since the group became established. Variation among species has, however, increased—that is, the range of brain sizes has grown as species become more numerous and more diverse in their adaptations. If we focus only on extreme values, we see a general increase through time and assume some intrinsic and ineluctable value in growing braininess. If we consider variation, we see only an expansion in range through time (leading, of course, to larger extreme values), and we offer a different explanation based on the reasons for increased diversity.

Second, an example of decreasing extremes properly interpreted as declining variation: A characteristic pattern in the history of most marine invertebrates has been called "early experimentation and later standardization." When a new body plan first arises, evolution seems to explore all manner of twists, turns, and variations. A few work well, but most don't (see essay 16). Eventually, only a few survive. Echinoderms now come in five basic varieties (two kinds of starfish, sea urchins, sea cucumbers, and crinoids—an unfamiliar group, loosely resembling many-armed starfish on a stalk). But when echinoderms first evolved, they burst forth in an astonishing array of more than twenty basic groups, including some coiled like a spiral and others so bilaterally symmetrical that a few paleontologists have interpreted them as the ancestors of fish. Likewise, mollusks now exist as snails, clams, cephalopods (octopuses and their kin), and two or three other rare and unfamiliar groups. But they sported ten to fifteen other fundamental variations early in their history. This trend towards shaving and elimination of extremes is surely the more common in nature. When systems first arise, they probe all the limits of possibility. Many variations don't work; the best solutions emerge, and variation diminishes. As systems regularize, their variation decreases.

From this perspective, it occurred to me that we might be looking at the problem of .400 hitting the wrong way round. League-leading averages are extreme values within systems of variation. Perhaps their decrease through time simply records the standardization that affects so many systems as they stabilize—including life itself as stated above and developed in essay 16. When baseball was young, styles of play had not become sufficiently regular to foil the antics of the very best. Wee Willie Keeler could "hit 'em where they ain't" (and compile an average of .432 in 1897) because fielders didn't yet know where they should be. Slowly, players moved toward optimal methods of positioning, fielding, pitching, and batting—and variation inevitably declined. The best now met an opposition too finely honed to its own perfection to permit the extremes of achievement that characterized a more casual age. We cannot explain the decrease of high averages merely by arguing that managers invented relief pitching, while pitchers invented the slider —conventional explanations based on trends affecting high hitting considered as an independent phenomenon. Rather, the entire game sharpened its standards and narrowed its ranges of tolerance.

Thus I present my hypothesis: The disappearance of the .400 hitter (and the general decline of league-leading averages through time) is largely the result of a more general phenomenon—a decrease in the variation of batting averages as the game standardized its methods of play—and not an intrinsically driven trend warranting a special explanation in itself.

To test such a hypothesis, we need to examine changes through time in the difference between league-leading batting averages and the general average for all batters. This difference must decrease if I am right. But since my hypothesis involves an entire system of variation, then, somewhat paradoxically, we must also examine differences between *lowest* batting averages and the general average. Variation must decrease at both ends—that is, within the entire system. Both highest and lowest batting averages must converge toward the general league average.

I therefore reached for my trusty *Baseball Encyclopedia*, that *vade mecum* for all serious fans (though, at more than 2,000 pages, you can scarcely tote it with you). The encyclopedia reports league averages for each year and lists the five highest averages for players with enough official times at bat. Since high extremes fascinate us while low values are merely embarrassing, no listing of the lowest averages appears, and you have to make your way laboriously through the entire roster of players. For lowest averages, I found (for each league in each year) the five bottom scores for

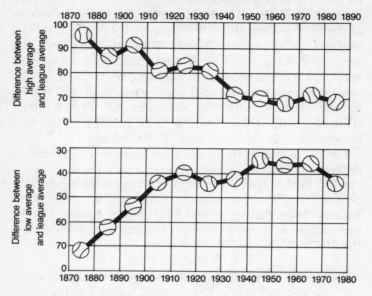

THE DECLINE IN EXTREMES

CREDIT: CATHY HALL.

players with at least 300 at bats. Then, for each year, I compared the league average with the average of the five highest and five lowest scores for regular players. Finally, I averaged these yearly values decade by decade.

In the accompanying chart, I present the results for both leagues combined—a clear confirmation of my hypothesis, since both highest and lowest averages converge towards the league average through time.

The measured decrease toward the mean for high averages seems to occur as three plateaus, with only limited variation within each plateau. During the nineteenth century (National League only; the American League was founded in 1901), the mean difference between highest and league average was 91 points (range of 87 to 95, by decade). From 1901 to 1930, it dipped to 81 (range of only 80 to 83), while for five decades since 1931, the difference between mean and extreme has averaged 69 (with a range of only 67 to 70). These three plateaus correspond to three marked eras of high hitting. The first includes the runaway averages of the 1890s, when Hugh Duffy reached .438 (in 1894) and all five leading players topped .400 in the same year (not surprising since that year featured the infamous experiment, quickly abandoned, of counting walks as hits). The second plateau includes all the lower scores of .400 batters in our century, with the exception of Ted Williams (Hornsby topped the charts at .424 in 1924). The third plateau records the extinction of .400 hitting.

Lowest averages show the same pattern of decreasing difference from the league average, with a precipitous decline by decade from 71 to 54 points during the nineteenth century, and two plateaus thereafter (from the mid-40s early in the century to the mid-30s later on), followed by the one exception to my pattern—a fallback to the 40s during the 1970s.

Patterns of change in the difference between highest and lowest averages and the general league average through time

	Difference between five highest and league average	Difference between five lowest and league average
1876–1880	95	71
1881–1890	89	62
1891–1900	91	54
1901–1910	80	45
1911–1920	83	39
1921–1930	81	45
1931–1940	70	44
1941–1950	69	35
1951–1960	67	36
1961–1970	70	36
1971–1980	68	45

Nineteenth-century values must be taken with a grain of salt, since rules of play were somewhat different then. During the 1870s, for example, schedules varied from 65 to 85 games per season (compared with 154 for most of our century and 162 more recently). With short seasons and fewer at bats, variation must increase, just as, in our own day, averages in June and July span a greater range than final-season averages, several hundred at bats later. (For these short seasons, I used two at bats per game as my criterion for inclusion in tabulations for low averages.) Still, by the 1890s, schedules had lengthened to 130–150 games per season, and comparisons with our own century become more meaningful.

I was rather surprised—and I promise readers that I am not rationalizing after the fact but acting on a prediction I made before I started calculating—that the pattern of decrease did not yield more exceptions during our last two decades, because baseball has experienced a profound destabilization of the sort that my calculations should reflect. After half a century of stable play with eight geographically stationary teams per league, the system finally broke in response to easier transportation and greater access to almighty dollars. Franchises began to move, and my beloved Dodgers and Giants abandoned New York in 1958. Then, in the early 1960s, both leagues expanded to ten teams, and, in 1969, to twelve teams in two divisions.

These expansions should have caused a reversal in patterns of decrease between extreme batting averages and league averages. Many less than adequate players became regulars and pulled low averages down (Marvelous Marv Throneberry is still reaping the benefits in Lite beer ads). League averages also declined, partly as a consequence of the same influx, and bottomed out in 1968 at .230 in the American League. (This trend was reversed by fiat in 1969 when the pitching mound was lowered and the strike zone diminished to give batters a better chance.) This lowering of league averages should also have increased the distance between high hitters and the league average (since the very best were not suffering a general decline in quality). Thus, I was surprised that an increase in the distance between league and lowest averages during the 1970s was the only result I could detect of this major destabilization.

As a nonplaying nonprofessional, I cannot pinpoint the changes that have caused the game to stabilize and the range of batting averages to decrease over time. But I can identify the general character of important influences. Traditional explanations that view the decline of high averages as an intrinsic trend must emphasize explicit inventions and innovations that discourage hitting—the introduction of relief pitching and more night games, for example. I do not deny that these factors have important effects, but if the decline has primarily been caused, as I propose, by a general decrease in variation of batting averages, then we must look to other kinds of influences.

We should concentrate on the increasing precision, regularity, and standardization of play—and we must search for ways that managers and players have discovered to remove the edge that truly excellent players once enjoyed. Baseball has become a science (in the vernacular sense of repetitious precision in execution). Outfielders practice for hours to hit the cutoff man. Positioning of fielders changes by the inning and man. Double plays are executed like awesome clockwork. Every pitch and swing is charted, and elaborate books are kept on the habits and personal weaknesses of each hitter. The "play" in play is gone.

When the world's tall ships graced our bicentennial in 1976, many people lamented their lost beauty and cited Masefield's sorrow that we would never "see such ships as those again." I harbor opposite feelings about the disappearance of .400 hitting. Giants have not ceded to mere mortals. I'll bet anything that Carew could match Keeler. Rather, the boundaries of baseball have been restricted and its edges smoothed. The game has achieved a grace and precision of execution that has, as one effect, eliminated the extreme achievements of early years. A game unmatched for style and detail has simply become more balanced and beautiful.

Postscript

Some readers have drawn the (quite unintended) inference from the preceding essay that I maintain a cynical or even dyspeptic attitude towards great achievement in sports—something for a distant past when true heroes could shine before play reached its almost mechanical optimality. But the quirkiness of great days and moments, lying within the domain of unpredictability, could never disappear even if plateaus of sustained achievement must draw in towards an unvarying average. As my tribute to the eternal possibility of transcendence, I submit this comment on the greatest moment of them all, published on the Op-Ed page of the *New York Times* on November 10, 1984.

STRIKE THREE FOR BABE

Tiny and perfunctory reminders often provoke floods of memory. I have just read a little notice, tucked away on the sports pages: "Babe Pinelli, long time major league umpire, died Monday at age 89 at a convalescent home near San Francisco."

What could be more elusive than perfection? And what would you rather be—the agent or the judge? Babe Pinelli was the umpire in baseball's unique episode of perfection when it mattered most. October 8, 1956. A perfect game in the World Series—and, coincidentally, Pinelli's last official game as arbiter. What a consummate swan song. Twenty-seven Brooks up; twenty-seven Bums down. And, since *single* acts of greatness are intrinsic spurs to democracy, the agent was a competent, but otherwise undistinguished Yankee pitcher, Don Larsen.

The dramatic end was all Pinelli's, and controversial ever since. Dale Mitchell, pinch hitting for Sal Maglie, was the twenty-seventh batter. With a count of 1 and 2, Larsen delivered one high and outside—close, but surely not, by its technical definition, a strike. Mitchell let the pitch go by, but Pinelli didn't hesitate. Up went the right arm for called strike three. Out went Yogi Berra from behind the plate, nearly tackling Larsen in a frontal jump of joy. "Outside by a foot," groused Mitchell later. He exaggerated—for it was outside by only a few inches—but he was right. Babe Pinelli, however, was more right. A batter may not take a close pitch with so much on the line. Context matters. Truth is a circumstance, not a spot.

I was a junior at Jamaica High School. On that day, every teacher let us listen, even Mrs. B., our crusty old solid geometer (and, I guess in retrospect, a secret baseball fan). We reached Mrs. G., our even crustier French teacher, in the bottom of the seventh, and I was appointed to plead. "Ya gotta let us listen," I said, "it's never happened before." "Young man," she replied, "this class is a French class." Luckily, I sat in the back just in front of Bob Hacker (remember alphabetical seating?), a rabid Dodger fan with earphone and portable. Halfway through the period, following Pinelli's last strike, I felt a sepulchral tap and looked around. Hacker's face was ashen. "He did it—that bastard did it." I cheered loudly and threw my jacket high in the air. "Young man," said Mrs. G. from the side board, "I'm sure the verb *écrire* can't be that exciting." It cost me 10 points on my final grade, maybe admission to Harvard as well. I never experienced a moment of regret.

Truth is inflexible. Truth is inviolable. By long and recognized custom, by any concept of justice, Dale Mitchell had to swing at anything close. It was a strike—a strike high and outside. Babe Pinelli, umpiring his last game, ended with his finest, his most perceptive, his most truthful moment. Babe Pinelli, arbiter of history, walked into the locker room and cried.

Postpostscript

Funny business. I labored for three years to write a monograph on the evolution of Bermudian land snails, and only nine people have ever cited the resulting tome. I wrote these few hundred words in a quarter hour's flood of inspiration during an interminable round of speechmaking at my son's annual Little League banquet (good for something besides sliced turkey, I always thought)—and it has already received more commentary than most of my technical papers combined.

Some people misunderstood (I received a blistering letter from Babe Pinelli's pastor, virtually demanding a public retraction of my charge that the great ump had consciously lied, either for an early shower or a place in the sun). I received many more lovely letters, including one from Pinelli's grandson who reported that "Babe never had second thoughts about the call and wouldn't hear of any ridiculing." Right on. One particularly kind broadcaster dug out his old tape of the incident and played it for me over the phone—after noting that Mrs. G. had deprived me of the pleasure, and that I had never actually heard the great moment.

I have been both pleased and amused to learn that this commentary, intended only as a sweet memory for a single event, has been read and discussed in several high school and college ethics classes. Just for the record, therefore, please don't read the piece as an argument for mushy relativism in the search for truth. The narrowly empirical issue has a clear and unambiguous factual resolution—an absolutely inviolable truth, if you will. The pitch was high and outside. Flexibility based on circumstance arises only with respect to definitions, invented by people and not part of the external world. The pitch, in that particular context, was a strike, and Pinelli was right.

I must also confess a profound embarrassment, especially in the light of my last paragraph. My original piece identified the pitch as *low* and outside (as reported by Peter Golenbock in *Dynasty*, his history of the Yankee glory years —but no excuses, I shouldn't have just copied). The *Times* even exacerbated the error by using as a title, not my intended line (now restored), but "the strike that was low and outside." But even error can have its reward, thus proving that the world contains some intrinsic benevolence. Red Barber, that fine man and greatest announcer of them all, corrected me ever so gently on his weekly five-minute gem for public radio. He oughta know; he was there after all (and

I wasn't, as the piece testifies). I checked far and wide just to make sure. He was right, of course. The pitch was high, not low. Remember that old cartoon series—"the thrill that comes once in a lifetime" (like the kid who drives his wagon up to the gas pump and says "fill 'er up"). That's how I felt. Can you just imagine—to be corrected by Old Redhead himself!

15 | Death and Transfiguration

TO MANY OUTSIDERS, Indianapolis is nothing but one weekend a year and 500 miles of auto racing. In continuous reality, it is an attractive city filled with modern amenities and a liberal sprinkling of those older structures that unite our frenetic and uncertain present with a more comforting past. Last week, on a break from stated duties, I wandered through the Murat Temple of the Shrine and the enormous cathedral of Scottish Rite Masonry. These lodges must once have dominated the social life of Indianapolis; they may yet, for all I know, be important. But their gigantic buildings look forlorn and abandoned—cavernous Victorian rooms in dark wood and stained glass, dimly lit by available light, filled with old, overstuffed chairs occupied rarely by a few elderly men in odd-shaped hats. Surely, the old order changeth.

I was in Indianapolis to attend the annual meeting of the Geological Society of America.* There I watched, listened, and joined the debate as a group of my colleagues in paleontology began to dismantle an old order of thinking about old objects—and to construct a new and striking approach to a major feature of life's history on earth: mass extinctions.

Paleontologists have known about mass extinctions from

*I wrote this essay in November, 1983—just after the meeting here described.

the inception of our science as a modern discipline. We have used them to mark the major divisions of our geological time scale—the boundaries between eras. The Permian extinction that rang out the Paleozoic era eliminated half the families of marine invertebrates; the Cretaceous extinction, marking the transition from Mesozoic to Cenozoic eras, wiped out some 15 percent of marine families, along with the most popular of all terrestrial creatures, the dinosaurs.

Nonetheless, though we have always acknowledged the reality of these great dyings, we have tried, in a curious way, to mitigate their effects, probably because our strong biases for gradual and continuous change force us to view mass extinctions as anomalous and threatening. We have, in short, attempted to depict mass extinction as a simple, quantitative extension of the slower disappearance, species by species, that characterizes normal times—larger and more abrupt, to be sure, but basically just more of the same. We have pursued two principal strategies to temper mass extinctions and bring them into harmony with events of ordinary times. First, we have emphasized continuity across the boundaries by trying to find direct ancestors for new forms that appear after an extinction among species that flourished just before the event. Second, we have toted the numerical patterns of extinctions to argue that the peaks were neither high nor abrupt enough to support a catastrophic view—in other words, we have argued that pulses of extinction were preceded by gradual declines lasting for millions of years, and that the peaks themselves do not stand so noticeably above the "background" rates of normal times.

Both these traditions were strongly challenged in Indianapolis in a series of separate and ostensibly unconnected papers that point to a common conclusion: mass extinctions must, by four criteria, be reinterpreted as ruptures, not the high points of continua. They are more *frequent,* more *rapid,* more *profound* (in numbers eliminated), and more *different* (in effect versus the patterns of normal times) than we had ever suspected. Any adequate theory of life's history will have to treat them as special controlling events in their own right. They will not be fully explained by the evolutionary theory we have constructed for interaction among organisms and populations of normal times—that is, by nearly all conventional evolutionary theory now available.

Adolf Seilacher, professor of geology at Tübingen in Germany, presented the centerpiece of this unplanned assault upon tradition. Dolf is the greatest observer I have ever had the privilege of knowing. He looks at common objects, scrutinized by generations of researchers, and invariably sees something new and unexpected. This time he turned his superior gaze upon the oldest of all metazoan (multicellular animal) assemblages—the Ediacaran fauna. His paper offered a fundamental reinterpretation of these fossils, complete with wide-ranging implications for the entire history of life—and I sat spellbound as wave after wave of expanded meaning cascaded over me.

About 570 million years ago, our modern fossil record began with the greatest of geological bangs—the Cambrian explosion. Within a few million years, nearly all major groups of invertebrates with hard parts made their first appearance in the fossil record. For fully three billion years before, life had included little more than a long sequence of bacteria and blue-green algae. But the fossil record of early life does include one important, if last-minute, exception—first discovered in Australia but now known throughout the world—the Ediacaran fauna (named for the major Australian locality). In rocks just predating the Cambrian explosion, we find a moderately diverse assemblage of medium to large (up to a meter in length), soft-bodied, shallow-water marine invertebrates.

In the continuationist tradition that I identified above as a first strategy for softening the impact of mass extinctions, paleontologists have constantly tried to identify the Ediacaran animals with modern groups. Thus, the Ediacaran animals have been interpreted as jellyfish, corals, and worms—a continuity of evolutionary relationship across the greatest of all geological boundaries. Yet, as I argue in the fol-

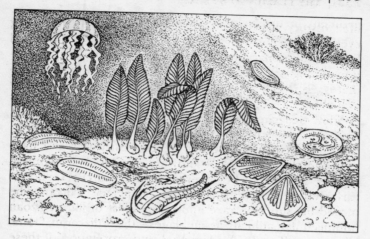

A conventional reconstruction of Ediacaran animals—my own, I regret to say. They are depicted as the ancestors of modern forms —jellyfish, soft corals, and worms. FROM *A View of Life* BY S.E. LURIA, S.J. GOULD, AND S. SINGER. AFTER AN ILLUSTRATION IN *The History of Life* (1977) BY A.E. LEE MCALESTER; REPRINTED BY PERMISSION OF PRENTICE-HALL, INC., ENGLEWOOD CLIFFS, N.J.

lowing essay, the traditional ploy of forcing old and problematical fossils into modern taxonomic categories often fails badly. We must recognize that the early history of life should be studded with failed experiments—small groups that never achieved much diversity and bear only distant relationship with any modern animal. We might expect that our oldest fauna should contain a large number of such curiosities—yet all Ediacaran animals have been shoehorned, often with considerable effort, into modern groups.

Dolf Seilacher now argues, turning the old view completely on its head, that the Ediacaran fauna contains no ancestors for modern organisms, and that every Ediacaran animal shares a basic mode of organization quite distinct from the architecture of living groups. The entire Ediacaran fauna, in other words, represents a unique and extinct experiment in the basic construction of living things. Our planet's first fauna was replaced after a mass extinction, not simply improved and expanded.

Dolf began by showing that the traditional similarities of Ediacaran and modern animals are misleading and superficial, and that the Ediacaran forms could not work as their supposed living counterparts. Nearly all Ediacaran fossils have been falsely fit into three modern groups: jellyfish, corals, and segmented worms. Living jellyfish move by contracting a prominent ring of concentric muscles located at the outer edge of their bell; radial grooves for gathering and transporting food lie within the concentric muscles, toward the center. But the so-called Ediacaran medusoids reverse this arrangement and therefore could not work in the same way: concentric structures surround the center, and radial grooves lie on the outside.

Modern alcyonarian corals ("soft" corals, or sea pens) invariably bear distinct branches, often springing from a common stem. The branches must be separated so that water, bringing oxygen and nutrients, can reach the individual polyps (members of the colony) growing on them. At first glance, the Ediacaran "sea pens" look superficially like their modern counterparts in general shape, but they form a continuous, quilted structure, not a set of separated branches—and could therefore not operate like a modern soft coral colony. The Ediacaran "worms" are segmented and bilaterally symmetrical like their supposed modern descendants but many other creatures share the same symmetry—and such a basic and repeatable architecture need not imply close relationship. In other respects, the Ediacaran creatures are most unwormlike. They stretch up to a meter in length and remain flat as a pancake—more like films than the substantially thickened bodies of most modern segmented worms.

After exposing the differences between Ediacaran animals and their supposed modern counterparts, Seilacher examined the similarities that unite all Ediacaran forms. They share an architecture only rarely used by modern animals—and not by any living creature ever linked to an

Ediacaran fossil. They look like ribbons, pancakes, and films, sometimes slightly "blown up" as air mattresses with a foliate or quilted structure.

The Ediacaran animals evolved before any creature had invented mineralized skeletons or external hard parts. Perhaps their unique *Bauplan* (to use the convenient German term for a basic scheme of organic architecture), records a pathway to large size that animals without supporting hard parts might follow—light and thin structures, woven together for added strength. In any case, and following a favorite theme of these essays for more than a decade, the Ediacaran fossils seem to represent one of two possible solutions—the one *not* followed by modern animals—to the basic structural problem of large size: the imposed decline of relative surface area since surfaces (growing as length squared) must increase more slowly than volumes (increasing as length cubed) as objects of similar shape get bigger. Since so many organic functions depend upon surfaces (respiration and feeding, to name just two) yet must serve the entire body's volume, such a decline in relative surface cannot be tolerated for long.

Of the two possible solutions, nearly all large modern animals have retained their rounded or globular shapes but have evolved internal organs to increase surface areas—the rich branching of airways in our lung, and the complexly folded surface of our small intestine, for example. Another potential solution, followed rarely today but exploited by some large parasites, including tapeworms, permits large size without internal complexity by changing the body's basic shape into something very thin—a ribbon or pancake—so that no internal space will be far from the external surface, the only site for respiration and absorption of food in the absence of internal organs. The Ediacaran animals, as a group, followed this second pathway to large size and therefore represent a coherent fauna strikingly different from any modern creature in basic design.

I might, were I inclined to search for progress in history, be gratified that life's first "try" used the simpler of two solutions—a change in body shape rather than an invention of complex internal organs. However, the more important point remains that if Seilacher is right, the Ediacaran fauna represents a different, unique, and coherent experiment in organic architecture—not a set of precursors for modern animals. To emphasize this discontinuity, the first Paleozoic fauna with hard parts, the so-called Tommotian assemblage, is filled with tiny tubed, coiled, and cap-shaped creatures bearing precious little resemblance to Ediacaran forms. The ancestry of these later creatures may be recorded in indirect evidence for other Precambrian animals not included among the Ediacaran fossils. We have found abundant remains, in "trace fossils" of tubes for feeding and burrowing but, alas, no "body fossils" as yet, of animals with more conventional rounded shapes—a good source for later Tommotian descendants.

Seilacher ended his paper with a particularly arresting argument. We have, he pointed out, been searching with no success, and little hope, for complex extraterrestrial creatures, primarily because we wonder so powerfully what an independent experiment in the development of life might produce. What similarities would another "try" show with life on earth? How strong a constraint do the physics and chemistry of objects impose? How different could life be elsewhere? Our answers may lie in the concrete evidence of our own fossil record, and not in the abstract speculations of exobiology. Perhaps an independent experiment occurred right here on earth, expressing itself as the Ediacaran fauna, our first assemblage of multicellular animals.

Returning to the theme of mass extinctions, we used to argue that the first era boundary, between Precambrian and Paleozoic some 570 million years ago, presented a puzzling difference from all others because it recorded a profound radiation (the Cambrian explosion) but no previous extinction. But if the Ediacaran fauna, lying just below the base of the Paleozoic in strata throughout the world, represents a coherent and different experiment in life's architecture, then a major extinction marks this initial boundary as well. The first strategy for mitigating mass extinction fails, and we trace little continuity across the opening and most profound boundary of life's complex history.

Other papers at Indianapolis challenged the second strategy by arguing for a greater separation in effect and magnitude between mass extinctions and events of ordinary times. Some conclusions of previous years, already documented in these essays, have paved the way: (1) An asteroidal impact as the source, or at least the *coup de grâce,* of our terminal Cretaceous extinction (essay 25 in *Hen's Teeth and Horse's Toes*)—organisms, after all, can scarcely "prepare" for such a trigger. (2) David Raup's estimate (essay 26 in *Hen's Teeth and Horse's Toes*) that a 50 percent extirpation of families, the counted figure for the Permian extinction, might translate to as much as 96 percent of all species (a removal of half the families implies an extinction of many more species since most species die without eliminating their families—a more inclusive category—while the death of a family must include all its species). For a removal so profound, we must seriously consider the possibility that entire groups will be lost for purely random reasons. (3) The calculation of Raup and Jack Sepkoski (essay 27 in *Hen's Teeth and Horse's Toes*) that major extinctions stand higher and more distinctly above the background level than previously recognized.

This theme of greater difference between mass extinctions and "normal" times gained strength and refinement in several papers presented at Indianapolis. Jack Sepkoski, a former student of mine now flourishing mightily at the University of Chicago, has spent years compiling the most consistent and complete data set ever developed for extinctions—a listing at the family level that includes everything from protozoans to mammals. With these data, we have finally achieved a basis for the fine-scaled consideration of quantitative patterns in extinction that this second strategy demands. (Good science may require genius and imagination, as these essays so often emphasize, but never forget that new conclusions are the fruit of hard empirical labor as well—otherwise, highfalutin thought is so much waffling.)

Using the Sepkoski data, Raup and Sepkoski have now identified a striking cyclicity in mass extinctions for 225 million years since the great Permian dying. Every 26 million years, with eight hits and just two apparent misses (a pattern too regular and striking to dismiss as accidental on statistical grounds), we find a peak of mass extinction; all previously identified disasters lie right on the highs of this 26-million-year cycle. What cause could yield a periodicity so regular, yet so widely spaced? If we understand geology aright, no purely internal process of climate, volcanism, or plate tectonics cycles so regularly with such a long period. Raup and Sepkoski therefore speculate that some astronomical cycle must be implicated—a solar or galactic phenomenon, although for the moment, we have no idea what.* If disasters are so frequent and caused by events so utterly beyond an organism's control or anticipation (how can populations track a 26-million-year cycle?), and if these coordinated dyings shape life's pattern so fundamentally, then mass extinction is not ordinary death extrapolated.

David Jablonski, a paleobiologist from the University of Arizona at Tucson, then added two cogent arguments to emphasize the abruptness and the different character of mass extinctions. For abruptness, Jablonski noted that the raw data of mass extinctions often include a long period of apparently slow and steady decline among groups that crash more profoundly at the peak itself. These slow declines have long been interpreted as a sign of continuity between normal and mass extinction. But are they real or an artifact of our imperfect geological record?

For more than one hundred years, geologists have sought terrestrial agents to associate with mass extinction. The litany is long, yet all but one have failed—mountain building, volcanism, fluctuations in temperature, to name just a few old and unsuccessful favorites. Falling sea level provides the one good correlation (and the 26-million-year-cycle theorists had better take it into account). Most mass

---

*See essay 30 for further, and rather remarkable, developments.

extinctions are preceded by a marked regression of sea level.

Falling sea level may participate as a cause of extinction (our fossil record is strongly biased toward shallow-water marine invertebrates), but it also imposes an obvious artifact upon our data. As sea level falls, fewer sedimentary rocks form to hold the fossils of limited oceans. Perhaps the slow decline that precedes most mass extinctions only records the decreasing volume of rock available for finding fossils, not a true and gradual decrease presaging the later peak.

Jablonski devised a clever method to measure the potential artifact. Some forms disappear from the record as sea level falls, only to come back again when seas return to deposit more rocks after the mass extinction itself. These temporary losses must record an artificial effect of falling seas and decreasing amounts of fossiliferous rock. Jablonski refers to these reappearing groups as "Lazarus taxa."

By counting the number of Lazarus taxa that disappear before, but reappear after, a mass extinction, Jablonski can estimate how much of a measured slow decline before a mass extinction might be the artificial result of less available rock for finding fossils, and how much must record a real and gradual event connecting peaks of mass extinction with normal times before.

In some cases, subtraction of the Lazarus taxa still leaves a residue of slow disappearance, and the pattern must be real (decline of ammonities before the Cretaceous extinction, for example). But for many Cretaceous groups, a measured slow decline can be attributed entirely to the artifact of decreasing available rock. Thus, the Cretaceous extinction, and others as well, may be more abrupt than we have previously realized. The case for an extraterrestrial agent gains strength. Mass extinction is something quick and special.

Jablonski then examined the behavior of groups during normal times and during episodes of mass extinction to see if he could detect consistent differences that might accentuate the special character of mass extinctions. He found some intriguing disparities. Some branches of the evolutionary tree contain many species either because new species form easily or because they resist extinction once they arise. Jablonski calls these branches "species-rich clades" as opposed to "species-poor clades," or groups that never contain many species.

During normal times, species-rich clades tend to increase their numbers of species continually—and to win increasing numerical advantage over species-poor clades. The environments of normal times must encourage either rapid speciation or persistence thereafter. But why, then, don't species-rich clades take over the biosphere entirely? Jablonski finds that these same species-rich clades fare worse than species-poor clades during mass extinctions. The individual species in species-poor clades have wider geographic ranges and broader ecological tolerances than the narrow-niched taxa of species-rich clades. This geographic and ecological breadth probably protects such species in the extreme environments that mass extinction must generate. These same features of breadth may cut down the rate of speciation in normal times (fewer opportunities for isolation and exploitation of new environments), thus rendering such groups species-poor.

This contrary behavior of species-rich clades in normal and catastrophic times preserves a balance that permits both species-rich and species-poor clades to flourish throughout life's history. More important in our context, this distinction emphasizes the qualitative difference between normal times and catastrophic zaps. Mass extinctions are not simply more of the same. They affect various elements of the biosphere in a distinctive manner, quite different from the patterns of normal times.

As we survey the history of life since the inception of multicellular complexity in Ediacaran times (see essay 16), one feature stands out as most puzzling—the lack of clear order and progress through time among marine invertebrate faunas. We can tell tales of improvement for some groups, but in honest moments we must admit that the history of complex life is more a story of multifarious variation about a set of basic designs than a saga of accumulating excellence. The eyes of early trilobites, for example, have never been exceeded for complexity or acuity by later arthropods. Why do we not find this expected order?

Perhaps the expectation itself is faulty, a product of pervasive, progressivist bias in Western thought and never a prediction of evolutionary theory. Yet, if natural selection rules the world of life, we should detect some fitful accumulation of better and more complex design through time—amidst all the fluctuations and backings and forthings that must characterize a process primarily devoted to constructing a better fit between organisms and changing local environments. Darwin certainly anticipated such progress when he wrote:

> The inhabitants of each successive period in the world's history have beaten their predecessors in the race for life, and are, insofar, higher in the scale of nature; and this may account for that vague yet ill-defined sentiment, felt by many paleontologists, that organization on the whole has progressed.

I regard the failure to find a clear "vector of progress" in life's history as the most puzzling fact of the fossil record. But I also believe that we are now on the verge of a solution, thanks to a better understanding of evolution in *both* normal and catastrophic times.

I have devoted the last ten years of my professional life in paleontology to constructing an unorthodox theory for explaining the lack of expected patterns during normal times—the theory of punctuated equilibrium. Niles Eldredge and I, the perpetrators of this particularly uneuphonious name, argue that the pattern of normal times is not a tale of continuous adaptive improvement within lineages. Rather, species form rapidly in geological perspective (thousands of years) and tend to remain highly stable for millions of years thereafter. Evolutionary success must be assessed among species themselves, not at the traditional Darwinian level of struggling organisms within populations. The reasons that species succeed are many and varied— high rates of speciation and strong resistance to extinction, for example—and often involve no reference to traditional expectations for improvement in morphological design. If punctuated equilibrium dominates the pattern of normal times, then we have come a long way toward understanding the curiously fluctuating directions of life's history. Until recently, I suspected that punctuated equilibrium might resolve the dilemma of progress all by itself.

I now realize that the fluctuating pattern must be constructed by a complex and fascinating interaction of two distinct tiers of explanation—punctuated equilibrium for normal times, and the different effects produced by separate processes of mass extinction. Whatever accumulates by punctuated equilibrium (or by other processes) in normal times can be broken up, dismantled, reset, and dispersed by mass extinction. If punctuated equilibrium upset traditional expectations (and did it ever!), mass extinction is far worse. Organisms cannot track or anticipate the environmental triggers of mass extinction. No matter how well they adapt to environmental ranges of normal times, they must take their chances in catastrophic moments. And if extinctions can demolish more than 90 percent of all species, then we must be losing groups forever by pure bad luck among a few clinging survivors designed for another world.

Heretofore, we have thrown up our hands in frustration at the lack of expected pattern in life's history—or we have sought to impose a pattern that we hoped to find on a world that does not really acquiesce. Perhaps now we can navigate between a Scylla of despair and a Charybdis of comforting unreality. If we can develop a general theory of mass extinction, we may finally understand why life has thwarted our expectations—and we may even extract an unexpected kind of pattern from apparent chaos. The fast track of an extraordinary meeting in Indianapolis may be pointing the way.

Postscript

As a happy irony of science at its best, any essay on exciting new material guarantees its own swift oblivion as discovery augments. I almost eliminated this essay as superseded (as others, not lamented, have disappeared), but finally decided to keep it without change as an honest expression of immediate excitement written while all the new ideas still rang in my ears. Thus, I have not tried to revise (and change the tone) with published versions that have appeared since the original verbal presentations. The essays of section 8 update the second part on mass extinction, while Seilacher's bibliographic reference may be consulted for more information on the first part.

I cannot, however, resist one update in pictorial form. In December, 1984, Dolf Seilacher sent me the following copy of his first attempt to draw the entire Ediacaran fauna in the light of his new theory. No theme is more basic to this book, and to its convictions about the centrality of history, than the importance of taxonomy viewed, not as a neutral hatrack for the objective facts of nature, but as a theory that constrains and directs our thinking. Seilacher's figure stunned me with the particular joy of seeing something entirely new in familiar objects. All my professional life, I have viewed the Ediacaran organisms as ancestors of later, modern phyla. I have so classified them in my mind. *Spriggina* (row 1) went with the worms, *Charnia* (row 1) with the corals, *Cyclomedusa* (row 3) with the jellyfish, and *Tribrachidium* (row 3) with the echinoderms. Placed in these disparate pigeonholes, I simply never saw the similarities that now jump out at me (although, in some "objective" sense, the similarities were always "there"). Now I can

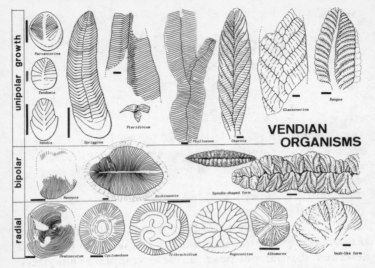

Seilacher's first drawing of the new taxonomy for Ediacaran organisms. Note how, arranged this way according to different styles and axes of growth—rather than by presumed relationship to later organisms—we can easily see the common features and coordinating themes of all Ediacaran organisms.

see Seilacher's point so clearly—a community of quilted, sheet-like structures with different axes of growth and symmetry. Taxonomy is a dynamic and creative science of history.

16 | Reducing Riddles

ON OCTOBER 1, 1939, a month after Stalin and Hitler signed their nonaggression pact, Winston Churchill described Russian policy as "a riddle wrapped in a mystery inside an enigma." All professions have their classical enigmas, although they can rarely boast a Churchill to describe them so well. My own field of invertebrate paleontology has a formal Latin designation for its mysteries. They are gathered into a wastebasket category of classification called Problematica—animals of unknown zoological affinity, even though their fossils may be both well preserved and abundant. The resolution of a problematic group becomes a cause for general rejoicing among paleontologists. Early in 1983, the most resolute of all paleontological mysteries finally yielded at least halfway. I wish to recount this tale and to explain why it has a general importance far transcending the simple delight of discovery.

The most vexatious of all fossil Problematica had been the conodonts (see photo on p. 246). As their name ("cone tooth") implies, conodonts are tiny, tooth-shaped structures of phosphatic composition. (Most hard parts of marine invertebrates are made of calcium carbonate, although some, including conodonts, are calcium phosphate. Vertebrate bone is also phosphatic, leading many paleontologists to speculate that conodonts might be the teeth of extinct fishes.) Conodonts range in size from microscopic dimen-

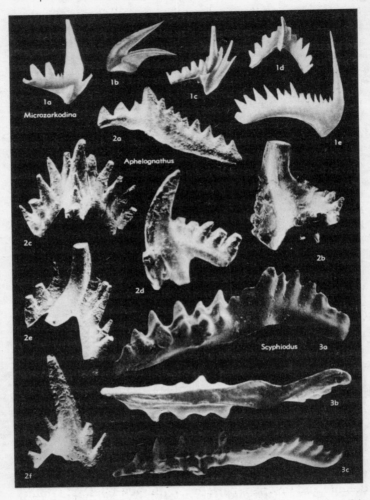

A selection of conodonts, tooth-like fossils of great stratigraphic value. REPRINTED FROM NATURAL HISTORY.

sions to about 3 mm in maximum length, and in age from Cambrian to Triassic—about 580 to 200 million years ago.

Many Problematica are rare and insignificant creatures. Conodonts, on the other hand (and despite their diminutive size), are among the most important of all fossils. They are found in abundance in a wide variety of rocks, and they evolved quickly, thus enhancing their value in correlating strata (since each short segment of time features its unique conodonts). Conodonts are therefore among the half dozen most important groups of fossils in the science of biostratigraphy—the dating and correlation of rocks by their fossil remains and still (despite growing interest in biological and evolutionary problems) the most important source of employment for paleontologists. One expert has stated that conodonts are "superb tools in worldwide biostratigraphy, and their value in Cambrian through Triassic rocks is not exceeded by any other group of fossils." Imagine, then, our frustration: such practical importance and we don't even know what kind of animal they represent. No other group of such importance lies among the Problematica.

Conodonts are evidently the only hard parts (and therefore the only portions generally preserved as fossils) of an otherwise soft-bodied creature. But what kind of animal, and how can you tell from a few separated toothlike structures? When conodonts were known only as isolated, disarticulated elements—the situation from their discovery in 1856 until 1934—we had almost no anchor for any sensible opinion, and speculation ran rampant. Conodonts were placed in almost every major group of the plant and animal kingdoms, from support structures for algae to copulatory organs of nematode worms. The most common opinions cast them as jaw elements either of annelid worms or of fishes.

In 1934, the first so-called assemblages of conodonts were discovered—articulated elements joined together in definite and invariant patterns. With their bilateral symmetry and gradation of toothlike elements from large to small, these assemblages suggested even more strongly that conodonts acted as food-gathering structures (either directly as teeth or indirectly as hard supports for fleshy or ciliary food collectors). More fanciful hypotheses of affinity faded away, and the idea that conodonts were jaw elements of some wormlike or fishlike creature gained further strength. But we still had no direct evidence for the conodont animal.

Then, in 1969, paleontologists throughout the continent gathered at the Field Museum of Natural History in Chicago for the first North American Paleontological Convention. (I well remember, as a wet-eared, first-year assistant professor, sitting there in awe amidst all the greats of my profession and thinking, "If the Russians—or the Chinese, or whoever—wanted to destroy this entire profession, one bomb. . . ." And then concluding confidently: "But why would they bother?") At the plenary session, a dramatic announcement was made—the conodont animal had finally been found. A soft-bodied creature from Montana had been discovered with conodonts inside, in a position interpreted as the mouth or anterior gut, where food might be chewed or macerated. These animals possessed other features that seemed to ally them with chordates, primitive members of our own phylum (including all vertebrates), and they were named conodontochordates.

It was, unfortunately, a false alarm. Further study revealed that the conodonts lay further back in the gut—in a position strongly implying that they had been swallowed by the beast. Moreover, their distribution was inconsistent with what we know about conodont assemblages. One conodontophore contained parts of two distinct assemblages, a clear indication that two conodont animals had somehow found their way within. Another contained conodonts varying too much in size for a reasonable inference that they came from the same organism. A third had no conodonts at all in the favored place. Clearly, the so-called conodontophores had been eating conodont animals and often retained conodonts of more than one individual in their gut. This news may have disappointed paleontologists, but it did not debase the significance of the discovery. The conodontophore is a conodont eater, not a conodont animal, but it remains an outstanding conundrum in its own right. Instead of resolving one of the Problematica, we had added another to our burgeoning list. So be it. The addition of one intriguing mystery is nearly as good (and often more interesting) as the solution to another.

Contrary to the romantic image of science and exploration, many important discoveries are made in museum drawers, not under adverse conditions in the parched Gobi or the freezing Antarctic. And so it must be, for the nineteenth century was the great age of collecting—and leading practitioners shoveled up material by the ton, dumped it in museum drawers, and never looked at it again. One of the great zoological discoveries of our century, the primitive segmented mollusk *Neopilina*, had been dredged from the deep sea, placed in a vial and labeled with the name of a limpet-shaped snail (for its external shell maintains such a shape)—where it remained for several years until H. Lemche turned the vial over to look at the soft parts and saw the segmented gills.

I am delighted to report that the conodont animal has now, and apparently truly this time, been found—in a museum drawer in Scotland. My friend Euan Clarkson was rummaging through some material of Carboniferous age (about 340 million years old) collected by D. Tait during the 1920s, when he noticed the impression of a worm-shaped creature with conodonts at the front end, right where the mouth should be. Since Clarkson is not an expert on conodonts, he called in some colleagues to verify and extend his discovery. Their results have just been published (Derek E.G. Briggs, Euan N.K. Clarkson, and Richard J. Aldridge, in bibliography).

Our fossil record is almost entirely the history of hard parts—bones, teeth, shells, and plates—because soft structures decay quickly and do not fossilize. Under very special circumstances, soft parts can be preserved, and these rare windows on the true diversity of past life are among the most precious of our fossil localities. For the 600 million years that multicellular animals have dominated our earth's fauna, we have no more than a dozen or so extensive deposits of soft-bodied creatures. Most famous are the carbonized films of weird and wonderful creatures from the Burgess Shale, Cambrian of Alberta (some 550 million years old, and most ancient of our extensive windows); animals preserved within ironstone concretions from the Mazon Creek

Formation of Illinois, Carboniferous period (350–270 million years old); and the Jurassic (180–130 million years) lithographic limestones of Solnhofen, Germany, where remains of *Archaeopteryx,* the first bird, feathers and all, were discovered.

The conodont animal comes from one of our smaller windows, the so-called shrimp band within the Granton Sandstones found east of Edinburgh. The Granton Sandstones are a sequence of lake and lagoonal sediments deposited in fresh or slightly saline water. This basin was occasionally flooded by the sea, and the shrimp band represents one such marine incursion. Its soft-bodied fauna was preserved because two unusual conditions prevailed during this brief flood. First, the bottom waters apparently lacked oxygen. No animal scavengers or bacteria could live on the lake floor, and dead animals floating down from above were not dismembered or decomposed. (We make these inferences because the shrimp band displays continuous, fine-layered sedimentation, an indication that no creatures burrowed or plowed through the bottom muck.) Second, the basin was stagnant and virtually devoid of currents. Thus, fragile, soft-bodied creatures were not pulled apart but floated gently down to be buried intact.

The conodont animal is wormlike in appearance, some 40.5 mm long, but no more than 2 mm wide (see photo on p. 251). Its head end seems to be cleft, with two broad lobes surrounding a central depression (entrance to the mouth, perhaps). Just behind the head, conodonts are affixed along one edge in a sensible position for the mouth. They occur in three groups and contain elements of a well-known assemblage. Thus, Clarkson and his colleagues did not need to invent a name for their creature; they included it within the genus *Clydagnathus,* first established in 1969 for the skeletonized conodonts alone. A few faint lines run along the interior of the animal, parallel to its sides. Whether these represent a gut, a nerve tube, perhaps even the chordate notochord, we do not know. About two-thirds of the way back, and extending nearly to the posterior end, we find an intriguing sequence of repeated segments, some thirty-

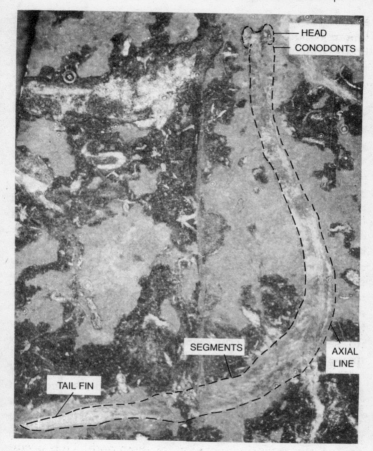

Fossil of a conodont animal, recently found in a museum drawer in Scotland. Its taxonomic position is controversial; it may best be placed in a new phylum of its own. REPRINTED FROM NATURAL HISTORY.

three in all, sloping at an angle to the midline of the body. Finally, one edge of the posterior end seems to sport a sequence of projections, interpreted as fin rays. Not much else worthy of mention has been preserved. At least the structures of *Clydagnathus* confirm one old assumption about conodont elements—they represent the only hard parts of an otherwise entirely soft-bodied creature. No wonder we had so little previous success in determining their affinity.

As I said at the outset, Clarkson and his colleagues have solved only half the conodont problem. They have found the elusive animal, but they do not know where it belongs. Of modern animal phyla, only two seem worthy of discussion as a possible taxonomic home for the conodont animal. Perhaps it is a chordate—that is, a prevertebrate member of our own phylum. Yet, each potential similarity with chordates scarcely carries conviction. The slender and flattened eel-shaped body reminds us of some chordates, but we find the same general shape in several other phyla as well. The faint lines parallel to the animal's side could represent such chordate structures as the notochord, but they may simply be remnants of the gut as well, an organ shared by virtually all "higher" animals. The apparent fin rays of the posterior end suggest chordate affinities, but similar structures occur in several other phyla as well. The V-shaped segments seem to say "chordate," but these structures are so poorly preserved that we cannot really distinguish between a chordate style of segmentation and the patterns of several other phyla with serially repeated elements. In short, we find a few general and superficial similarities with chordates but nothing specific, and surely nothing that would warrant any firm, or even tentative, placement within our phylum.

The Chaetognatha, or arrow worms, a small marine group located not far from chordates on our evolutionary tree, include the only other viable candidates for a link between the conodont animal and some modern group. Chaetognaths are armed with grasping spines that flank the mouth in two lateral sets. These spines bear a superficial resemblance to conodonts, but they are made of chitin, not calcium phosphate. Chaetognaths also have tail fins not unlike those of the conodont animal. But they also have lateral fins, and such structures are not present on the conodont animal (in an area of the body—the posterior—where preservation is detailed and excellent). In short, chaetognaths seem an even less worthy prospect than chordates as a home for the conodont animal.

Briggs, Clarkson, and Aldridge therefore conclude, with ample justice in my opinion, that the conodont animal is unique and previously unknown. It must be placed in a separate phylum—the Conodonta. After all, they argue, if a century of efforts to squeeze them into some modern group have been dashed on the enigma of their peculiar hard parts, why should the discovery of equally ambiguous soft parts comfortably fit them into some well-established pigeonhole of our taxonomy? They write: "The lack of a definitive solution to this problem in 125 years of research emphasizes the uniqueness of conodonts." And with this conclusion—that conodonts must be placed in a new and separate phylum of their own—we finally come to the general message that inspired me to write this essay.

Paleontologists are, in general, a conservative lot. Problematica of uncertain taxonomic affinity and few species are an embarrassment and an untidy bother; nothing makes an old-style paleontologist happier than the successful housing of problematical organisms within a well-known group. The admission that Problematica must be treated by erecting new phyla flies in the face of hope and tradition and represents a last resort. In recent years, that resort has been followed more and more often because—well, damn it—many Problematica are weird, wonderful, and unique and simply do not fit into any known group. This most unwilling admission reflects an important and little-known fact about the history of life.

To appreciate this fact and its implications, we must study the distribution in time of Problematica that cannot be placed into conventional phyla. The history of life has featured multicellular animals only during the past 600 million years. We divide this time into three great eras—the Paleozoic (or ancient life), the Mesozoic (or middle life), and the Cenozoic (or recent life). Virtually all Problematica now begrudgingly granted their own phylum lived during the oldest era, the Paleozoic (although conodonts, after living throughout the Paleozoic, just sneaked into the Triassic, the first period of the Mesozoic). This fact, the focal point of my essay, may not strike you, at first, as strange. After all, the further back we go, the more different should life become from our modern phyla. But two aspects of this distribution in time are surprising and point to a major pattern. First, although we might expect a general decrease in the number of problematic groups through time, we would not anticipate an abrupt disappearance of oddities after the Paleozoic. We do not find a gradual decline in curious creatures. Instead, they abound in the lower Paleozoic, become rare by the end of the Paleozoic, and cease thereafter. Of the three windows I mentioned, the Burgess Shale (lower Paleozoic) is chock-full of Problematica, the Mazon Creek (upper Paleozoic) sports two, the Solnhofen lithographic limestone (Mesozoic), none. Something about the earliest history of multicellular life encouraged a flowering of Problematica. Something about its later history (and not much later) dried the well completely.

Second—although the conodonts are an exception to this generality—most Problematica are rare, restricted in time, and represented by only a few species. Phyla are supposed to be big groups—arthropods with their 750,000 species of insects, or chordates with their 20,000 species of fishes. They are also supposed to endure for a long time. Taxonomists are stingy; they do not like to establish a group just below the highest level of kingdom for just a few species that lived but a few million years. If Problematica were restricted to the Paleozoic but were all as abundant and long-lived as conodonts, the pattern would not be as troubling or curious. But some Problematica, now housed in their own phylum, are known as only one species found in a single place. And some are surpassingly strange. Consider the animal so formidably curious that it goes by the Latin name *Hallucigenia,* coined by its author, Simon Conway Morris, for "the bizarre and dream-like appearance of the animal." (Simon once told me that it resembled something he had seen on a trip—and I don't mean to Boston.) *Halluci-*

genia (from that first and most famous window, the Burgess Shale) has an elongate body, nearly an inch in length, supported by seven pairs of spines that look nothing like the legs of any known creature. It has a bulbous head and, behind it, a row of tentacles, each forked at the tip, running along the back. Behind the tentacles lies a smaller and bunched array of projections recalling the spines on a *Stegosaurus*'s tail. An anal tube projects upward at the rear end (see figure on p. 256). Damned strangest thing I've ever seen in my life. Or consider the peculiar Problematica from the second window, our own Mazon Creek Formation of Illinois. It also bears a whimsical formal name, a Latinization of its discoverer, a Mr. Tully, and its appearance. It is called *Tullimonstrum*. The Tully monster is a peculiar, roughly banana-shaped creature, some three to six inches long. Like *Hallucigenia,* it is so different from anything else we know that it seems to demand a phylum of its own.

We tend to think of evolution as progressive change within lineages—fish become amphibians, reptiles, mammals, and finally humans—and we therefore miss important themes related to a different and more pervasive aspect of evolution: changing diversity, considered as absolute numbers of species and their relative abundance through time. The predominance of Paleozoic Problematica records an outstanding theme in the history of diversity. This theme imparts a direction to time that is more clear and reliable than any statement we can make about change within lineages. It also probably reflects a more general and basic law about the history of change in natural systems.

During the past decade, paleontologists have hotly debated the pattern of change through time in the diversity of marine animals. Do more species live now (as the "progressivist" view of evolution might suggest) or has the number of species remained roughly constant following a quick achievement of some equilibrium value after the Cambrian explosion? The problem is not so easy to solve as it may seem. You can't simply count the number of species described for each interval of time. The fossil record is notoriously imperfect, and it tends to get worse the further back

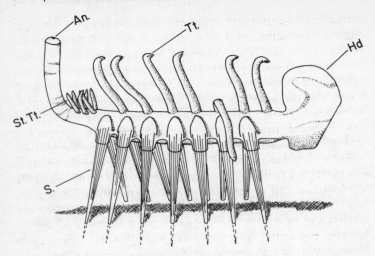

Simon Conway Morris's restoration of *Hallucigenia sparsa*. Note the seven pairs of spines below (labeled S), the bulbous "head" (in front, labeled Hd), the single row of forked tentacles on the back (labeled Tt), the bunched array of projections at the rear (labeled St. Tt.) and the upstanding "anal tube" (labeled An). FROM PALAE-ONTOLOGY, VOLUME 20, 1977, P. 628.

you go. Thus, an empirical increase in abundance of known fossils could actually reflect a decrease of true diversity.

Arguments have therefore raged back and forth, but in 1981 the four leading debaters buried the hatchet and published a joint paper of welcome agreement (J.J. Sepkoski, R.K. Bambach, D.M. Raup, and J.W. Valentine, in bibliography). Several sources of data (all corrected as best we can for imperfection of the record) now point to a clear pattern of real increase through time—not steady and progressive but unmistakable as a general direction. Modern oceans contain at least twice the number of species as our average Paleozoic seas.

We might therefore expect—indeed it seems unavoidable —that modern seas would not only contain more species but also more distinct kinds of creatures, more basically different body plans. Yet it is not so. Today, double the number of species are crammed into far fewer groups of higher taxonomic rank. To be sure, we still find several phyla of distinct body plan and low membership—all the wormlike groups with the funny names that no one but specialists know and love: the kinorhynchs, the gnathostomulids, the priapulids, the chaetognaths, already mentioned as a possible home for conodonts, and several others. But our modern seas are dominated by just a few groups—primarily clams, snails, crabs, fishes, and echinoids —each with far more species than any Paleozoic phylum ever attained (with the possible exception of trilobites in the Ordovician and crinoids in the Carboniferous). Paleozoic seas may have contained only half the species that grace our modern oceans, but these species were distributed over a greatly expanded range of basic body plans. This steady decrease in kinds of organic designs—all in the face of a strong increase in numbers of species—may represent the most outstanding trend of our fossil record.

This steady decrease is well recorded by the pattern of Problematica already discussed. Most of the really weird and wonderful creatures lived exclusively during the Paleozoic. (Don't be too impressed by the oddity of some modern minor phyla, for many of them did not arise recently but also have records extending back into the Paleozoic.) It is perhaps even better recorded by changes in the number of classes (next lower taxonomic level) within common phyla. Consider just one example, based on a highly conservative counting of classes made by J.J. Sepkoski of the University of Chicago. Modern echinoderms come in four classes, all of respectable to high diversity: sea urchins (the echinoids already cited as a dominant group), starfishes, sea cucumbers, and crinoids. Yet, sixteen additional classes lived and died within the Paleozoic, and sixteen of the total twenty coexisted during the Ordovician period, some 500 million years ago. None of these sixteen classes (with two possible exceptions) ever reached the diversity displayed today by any of the modern survivors.

The Paleozoic world was very different from ours, with few of a kind distributed over a greatly increased range of basic body forms. *Hallucigenia* is gone, the Tully monster lives no longer, even the abundant conodonts are extinct. Why has our world of life undergone this profound shift from few species in many groups to many species in fewer groups?

Of the two general answers, the first is conventional and causal (the second will be based on random processes). It invokes what may be a common property of nearly all natural systems and may therefore have an importance far transcending this particular example. The principle might be called "early experimentation and later standardization." Some 600 million years ago, the Cambrian explosion filled the oceans with their first suite of multicellular animals. Evolution probed all the limits of possibility. Each basic body plan experimented with a great array of potential variants. The pattern of many groups, each with few members, was established. Some of these experiments worked well, but, inevitably, most didn't—and a gradual sorting-out ensued.

Many of the failures were flawed from the start and never reached high diversity. They are our taxonomic embarrassments—highly distinct body plans with few species. We call them Problematica and grant them their own phyla only begrudgingly (although if we understood the principle that they represent, we would propose and accept their special names with more equanimity). Others, like the small and extinct classes of Paleozoic echinoderms, are failed experiments with a basic design that does work well in a few successful classes. Thus, sea urchins and starfishes use the echinoderm ground plan to great advantage, while a host of early experiments, bearing such strange names as ctenocystoids, helicoplacoids, and edrioblastoids, quickly bit the dust. Our modern faunas are the winnowed and well-honed survivors of a grand sorting-out based on principles of good engineering.

The same principle applies to any system free to experiment but ultimately regulated by good and workable design. Electric and steam cars, and a variety of other experiments, yielded to the internal combustion engine (although someday, if we ever run out of oil, they may reemerge like the phoenix). Cars now come in hundreds of brands, each built on the same principle. In 1900, far fewer brands used a much greater variety of basic designs. And consider the blimps, gliders, and variety of powered planes before we settled upon 747s and their ilk.

This principle of early experimentation and later standardization dictates a general reduction of variation—particularly the elimination of extremes. We often misunderstand the reason for a loss of extremes because we try to interpret the disappearance of oddities as a trend in its own right and not as an inevitable consequence of decreasing variation within a natural system. Essay 14 on the disappearance of .400 hitters in baseball considers another example of the same process. Conventional explanations for this most striking and widely discussed trend in baseball invariably look for some directional change—introduction of relief pitching or more grueling schedules composed mostly of night games—that would diminish high hitting alone. But I reasoned that the decline of high hitting may simply reflect the stabilization and general perfection of play that must accompany a game as its standards rise (analogous to the reduction of body plans as successful designs predominate in life's history). As pitching, fielding, and hitting all improve, variation in each category decreases. I was able to show that league averages have not changed between the great era of .400 hitting (1890–1920) and today, but that both highest averages (the .400 hitters) *and* lowest averages have converged toward the league average. In other words, extremes have been eliminated at both ends—the same principle of early experimentation (or toleration) and later standardization.

The second explanation is unconventional and based on random processes. A pattern of shift from few species in many groups to many species in fewer groups would occur even under regimes of random extinction, provided that we allow greater average change per event of speciation early in the history of life (as seems warranted in an initially "empty" world open to almost any experiment in form).

Extinction, as ecoactivists remind us, is forever. Once we lose a complex experiment in form, it will not arise again; the mathematical odds are too strongly against such a repetition of numerous complex steps (biologists refer to this principle as "the irreversibility of evolution"). Thus, inevitably, we lose most of the early experiments and begin to fill our oceans with repeated examples of the few surviving major groups. Intrigued as I am by random processes, I doubt that they will explain our entire pattern of reduction in body plans, if only because the idea of early experimentation and later standardization makes so much sense. But I would urge that the predictable consequences of random processes be granted much more attention than they usually receive. Random processes do produce high degrees of order—and the existence of pattern is no argument against randomness.

We live in a world of history and change. As creatures of habit who feel comforted by the discovery of order, we search for principles that grant time a direction—that admit a bit of order into the buzzing and blooming confusion of history. But arrows of time are hard to find and science hasn't given us many. The second law of thermodynamics, with its increasing entropy and decreasing order in closed systems, is our most famous agent of direction. Most proposals from evolutionary biology are spurious and based more on our hopes and expectations than the workings of natural selection—the notion of continual progress in particular. But this principle of diversity—early experimentation and later standardization—may be a true mark of history, producing trends towards decreased variation in basic designs of life. We should therefore care about conodonts, even if we have never correlated a rock or tend to look askance at inch-long worms with faint tail fins and bilobed heads. For their age, their taxonomic uniqueness, and their demise may record the nature of history.

5 | Politics and Progress

17 | To Show An Ape

TODAY, WE CLASSIFY all humans in a single species, *Homo sapiens*. But Carolus Linnaeus, in the founding document of animal taxonomy, the *Systema Naturae (System of Nature)* of 1758, recognized a second species, *Homo troglodytes*. While Linnaeus devoted several pages to *Homo sapiens* in all our diversity, *Homo troglodytes* merited only a paragraph. This second species, active only at night and speaking in hisses, offered little information to back up its existence. *Homo troglodytes* emerged as a compound of exaggerated travelers' reports based on imperfect observations of anthropoid apes humanized or native peoples degraded. Linnaeus even ventured the possibility of a third species, *Homo caudatus,* or man with a tail, but he admitted that this creature, *incola orbis antarctici* (an inhabitant of the antarctic regions), remained so obscure (if it existed at all) that he could not determine "whether it belongs to the human or monkey genus."

Why did this sober naturalist include such poorly supported fiction in the description of his first and most important genus? As a basic answer, Linnaeus worked with a theory that anticipated such creatures; when something should exist anyway, imperfect evidence becomes more acceptable.

I often write about the interaction of theory and fact in these essays because no other theme so well displays the human side of science—the intrusion of mind into nature and their necessary interpenetration in all creative activity. Science does not follow a one-way path from yielding nature to objective mind. This theme also illustrates why we must abandon as bankrupt the common procedure of judging past scientists by their accuracy according to present knowledge. Some incorrect theories, as grand and generous syntheses of knowledge, pose large and exciting questions, and may thereby produce as many new discoveries as notions that we accept today (see essay 6 in *Hen's Teeth and Horse's Toes* on James Hutton's use of final causes).

In this case, an incorrect theory, the chain of being, led Linnaeus to expect intermediate forms between apes and humans. For the objects of nature formed a single chain stretching without a break from the simplest amoeba to us. But the chain of being had always faced a substantial empirical problem—large and apparent gaps between major units, in particular, minerals and plants, plants and animals, monkeys and humans (see essay 18 for a further discussion of this problem). Indeed, Sir Thomas Browne, in his *Religio Medici* (1642), had declared that the gaps increased as we mounted the scale:

> There is in this Universe a Stair, or manifest Scale of creatures, rising not disorderly, but with a comely method and proportion. Between creatures of mere existence, and things of life, there is a large disproportion of nature; between plants and animals or creatures of sense, a wider difference; between them and Man, a far greater: and if the proportion hold on, between Man and Angels there should be yet a greater.

For those dedicated to filling the gaps, the apparent distance between monkey and human posed the greatest resolvable dilemma—and *Homo troglodytes* fit right in between.

But if *Homo troglodytes* only recorded the vivid imagination of early travelers, the great apes—gibbons, chimps, orangs, and gorillas—did exist. None was adequately known or described in Western Europe before the seventeenth century, thus increasing the apparent distance between humans and the most advanced primate. Arthur O. Lovejoy, in his classic treatise, *The Great Chain of Being,* explicitly cited the impetus given to the study of apes as an important empirical consequence of this false theory. He wrote:

> The principle of continuity was not barren of significant consequences. It set naturalists to looking for forms which would fill up the apparently "missing links" in the chain. . . . The metaphysical assumption thus furnished a program for scientific research. It was therefore highly stimulating to the work of the zoologist. . . . It therefore became the task of science at least to increase the *rapprochement* of man and ape.

The first adequate description of a great ape was not published until 1699, exactly one hundred years before the last great defense of the static chain—Charles White's treatise, analyzed in the next essay. In that year, Edward Tyson, England's finest comparative anatomist, published his "Orang-Outang, sive *Homo sylvestris:* or, the anatomy of a pygmie compared with that of a monkey, an ape, and a man." (In Tyson's day, orangutan, literally man of the forest, served as a general term for all great apes, both African and Asian, not only for the Asian form, as today. Tyson, overly cautious in this case, also doubted reports of Pygmy humans in Africa and assumed that his baby chimpanzee, which he falsely regarded as nearly full grown, formed the source of such rumors.)

Edward Tyson (1650–1708) studied at both Oxford and Cambridge, and then practiced as a physician in London. He taught human anatomy for fifteen years at Surgeon's Hall and became chief physician to England's most noted mental hospital, Bethlehem (whence our word *bedlam*). There he introduced the practice of female nursing and set up a department to follow patients after their release, an early example of outpatient care. He was, however, best known as a comparative anatomist and specialist on glandular systems. He wrote monographs on a porpoise and an

Orang-Outang, sive Homo Sylvestris:

OR, THE
ANATOMY
OF A
PYGMIE

Compared with that of a

Monkey, an *Ape,* and a *Man.*

To which is added, A

PHILOLOGICAL ESSAY
Concerning the
Pygmies, the *Cynocephali,* the *Satyrs,* and *Sphinges*
of the ANCIENTS.

Wherein it will appear that they are all either *APES* or
MONKEYS, and not *MEN,* as formerly pretended.

By *EDWARD TYSON* M. D.

Fellow of the Colledge of Physicians, and the Royal Society:
Physician to the Hospital of *Bethlem,* and Reader of
Anatomy at *Chirurgeons-Hall.*

LONDON:

Printed for *Thomas Bennet* at the *Half-Moon* in St. *Paul's* Church-yard;
and *Daniel Brown* at the *Black Swan* and *Bible* without *Temple-Bar*
and are to be had of Mr. *Hunt* at the Repository in *Gresham-Colledge.*
M DC XCIX.

The title page of Tyson's *Anatomy of a Pygmie* (1699).

opossum, but his 1699 treatise on a young chimpanzee became his most famous and enduring work. He was a wealthy, quiet, and conservative man who never married and showed unusual dedication to his anatomical studies and his avocation of classical learning. A funeral poem written to his memory in 1708 celebrated his sole devotion to Minerva, goddess of wisdom, among women:

> No Brow cou'd richer Chaplets ever twine,
> At least with Gems from Wisdom's sacred Mine.
> No Wonder ne'er by Beauty Captive led,
> No Bridal Partner ever shared his Bed.
> No, to the blinder God no Knee e'er paid,
> To great Minerva his whole Court he made.

Ashley Montagu's fine biography of Tyson (see bibliography) remains the standard work on this important but neglected figure in the history of science.

We now regard great apes as the most humanlike of primates and the closest to us in ancestry among living forms. However, great apes and humans differ substantially, not only in anatomy, but particularly in speech and mental functioning. Chimpanzees, our closest living relatives, are members of an evolutionary side branch, not ancestors or intermediate forms. But Tyson placed his pygmy, or juvenile chimpanzee, squarely in the middle between other primates and humans. When forced to categorize, Tyson did place his pygmy among the animals: "Our Pygmie has many advantages above the rest of its species, yet I still think it but a sort of ape and a mere brute; and as the proverb has it, an ape is an ape, tho' finely clad." Yet, in several other passages, Tyson demands an intermediate status for his chimp: "Our Pygmie is no man, nor yet the common ape; but a sort of animal between both." (In Tyson's day, before scientists had recognized the great apes as a separate group among primates, the term *ape* referred to any large monkey.)

Tyson's willingness to place great apes even nearer to humans than current understanding permits has become the source of a major historical misunderstanding about him—and the initial impetus for this essay based upon my continuing concern with the relationship between fact and theory. In the "heroic" school that analyzes past figures in terms of their success by modern standards, Tyson wins high acclaim for his courage in recognizing, so long ago, the affinity of apes and humans. He was able to discern this basic truth, the myth continues, for two major reasons: he was an outstanding empiricist, willing to cast aside old prejudices and simply record what he saw; and he used the modern method of comparative anatomy—explicit contrasts, part by part, of his chimp with other primates and humans.

This tradition of praising Tyson for his supposed modernism pervades the history of comment on his great 1699 treatise. T.H. Huxley, for example, in his seminal essay on *Man's Place in Nature* (1863), singled out Tyson for praise because he had written "the first account of a manlike ape which has any pretensions to a scientific accuracy and completeness." In the preface to his biography, Ashley Montagu states that he first became interested in Tyson when he read, as a student, the comment in a standard text on anthropology (1904) that Tyson's work "constitutes a most remarkable anticipation of modern methods of research." George Sarton, our century's foremost historian of science, wrote in the foreword to Ashley Montagu's biography that Tyson's treatise "is one of the outstanding landmarks in the history of science . . . a landmark in the history of the theory of evolution"—even though Tyson speaks only of the static chain, not of evolution at all.*

*D.J. Boorstin's wonderful book, *The Discoverers* (New York, Random House, 1983), has been published since this essay appeared. It continues the unfortunate tradition of praising Tyson as a courageous modernist and harbinger of evolution, not realizing that his discovery of intermediacy did not foment a revolution, but rather solved a problem in the standard "chain of being" theory as understood in Tyson's own time. Boorstin writes (p. 461): "Just as Copernicus displaced the earth from the center of the universe, so Tyson removed man from his unique role above and apart from all the rest of Creation. . . . Never before had there been so circum-

The myth of Tyson's supposed courageous modernism is belied by two anomalies, also prominently reported. First, if he was such an iconoclast in his willingness to place an animal so near our exalted selves, why is he universally described as so cautious and conservative in character? Secondly, if the award of intermediate status to his chimp was so controversial, why did it elicit so little contemporary comment—even though later generations granted Tyson so much praise? Ashley Montagu states: "The fact that Tyson is so little referred to in contemporary correspondence is not a little puzzling."

I believe that the solution to this dilemma lies simply in abandoning the fallacious approach to the history of science that generated it. Tyson was no modernist. He was a conservative man and he worked under the common preconceptions of his time. He did not place his chimp in an intermediate position between monkeys and humans because he anticipated evolution or simply saw more clearly through a veil of common prejudice. Rather, Tyson was a staunch exponent of the chain of being—a common and accepted ordering of nature in his time. Gaps between major groups vexed this theory sorely—and the space between monkey and man seemed particularly glaring and embarrassing. Scientists sought intermediate forms eagerly (and anxiously); Tyson's discovery produced a welcome confirmation of an established theory—the static chain of being—not a challenge based on a radically different idea—evolution— that would not be widely and seriously discussed for another century. Tyson's work received little comment because it was comforting and noncontroversial.

Moreover, Tyson's use of the comparative method does not mark him as an enlightened modernist, but also arises from his commitment to the chain of being. If you wish to place an animal between a monkey and a human, what else

stantial or so public a demonstration of man's physical kinship with the animals. . . . The implication was plain that here was the 'missing link' between man and the whole 'lower' animal creation. . . . Just as the heliocentric vista once seen could not be forgotten, so, after reading Tyson, who could believe that man was an isolate from the rest of nature."

can you do but tabulate its relative resemblance to each?

I do not in the least mean to criticize Tyson or to detract from his legitimate place in the pantheon of scientific heroes. Fitting a man into his time should only enhance our understanding. After reading Tyson's treatise, I can certainly affirm the elaborate care and accuracy of his descriptions, attributes highly valued in any age. Still, as the major theme of this essay, I want to argue that the outstanding feature of Tyson's treatise is not an accuracy emerging from the renunciation of old prejudices, but rather Tyson's *exaggeration* of the humanlike character of his pygmy—a result of his prior commitment to the chain of being. Theory always influences perception, and not always for the worse.

Tyson states right at the outset his commitment to the chain of being and his intention to use it as the organizing theme of his treatise.

> 'Tis a true remark, which we cannot make without admiration, that from minerals, to plants; from plants, to animals; and from animals, to men: the transition is so gradual, that there appears a very great similitude, as well between the meanest plant, and some minerals; as between the lowest rank of men, and the highest kind of animals. The animal of which I have given the anatomy, coming nearest to mankind, seems the nexus of the animal and rational.

He then defends the comparative technique, not as something controversial and modern, but as the appropriate method for placing a creature in the scale of being:

> To render this disquisition more useful, I have made a comparative survey of this animal, with a monkey, an ape, and a man. By viewing the same parts of all these together, we may better observe nature's gradation in the formation of animal bodies, and the transitions made from one to another; than which, nothing can more conduce to the attainment of the true knowledge, both of the fabrick, and uses of the parts. By

following nature's clew in this wonderful labyrinth of the creation, we may be more easily admitted into her secret recesses, which thread if we miss, we must needs err and be bewilder'd.

Despite several assertions that, at what later centuries would call the "bottom line," his pygmy was a "brute" and not a rational creature, Tyson continually emphasizes the humanlike qualities of his chimpanzee. At the very end, in a list of features, he cites forty-eight points of greater resemblance between chimp and human than chimp and ape, and only thirty-four for closer affinity between chimp and ape. The entire text continually emphasizes the smoothly intermediate position of Tyson's chimp: "In this chain of the creation, as an intermediate link between an ape and a man, I would place our pygmie."

Since chimps are, in general anatomical aspect, probably more similar to other primates than to humans, this conclusion requires some exaggeration of the humanlike qualities of Tyson's pygmy. Quite unconsciously, I suspect, and for two quite different reasons, Tyson continually overemphasizes the human similarities and as often underestimates the relationship with apes.

For the first reason, Tyson simply and consistently favors the human side in ambiguous situations. Note particularly his statements on posture. Tyson's chimp was brought to England from Angola and arrived both ill and very weak (it died within a few months and thus became available for Tyson's dissection). He observed that it occasionally but rarely walked erect; Tyson's chimp usually progressed, as great apes characteristically do, by walking on its knuckles —feet firmly on the ground but hands bent over. Tyson attributed this peculiar posture to its weakened state and insisted that its natural mode of locomotion must be erect on legs alone, as in humans—even though his empirical data identified knuckle walking as far more common:

When it went as a quadruped on all four, 'twas ackwardly; not placing the palm of the hand flat to the ground, but it walk'd upon it's knuckles, as I observed it to do, when weak, and had not strength enough to support it's body. . . . Walking on it's knuckles, as our pygmie did, seems no natural posture; and 'tis sufficiently provided in all respects to walk erect.

We can't blame Tyson for not knowing that great apes normally walk on their knuckles, for this most uncharacteristic pose for animals was not well described in his time. Still Tyson's defense of upright (or humanlike) posture as the normal mode for chimps does seem a bit forced, conditioned more by preconceptions about intermediate status on the chain of being than by direct attention to raw data. Thus, in writing "we may safely conclude, that nature intended it a biped," Tyson discusses the articulation of femur to pelvis and the "largeness of the heelbone in the foot, which being so much extended, sufficiently secures the body from falling backwards." Yet in the same discussion he conveniently omits other anatomical features described earlier that might lead us to doubt upright posture—particularly the major differences in pelvic structure between chimps and humans, and the handlike foot with its short and weak big toe.

Since primates are visual animals, we must never omit (though historians often do) the role played by scientific illustrations in the formation of concepts and support of arguments. Tyson's magnificent plates are all constructed to enhance the argument for upright posture, even in the absence of direct evidence for it (I include four reproductions with this essay). The first shows his pygmy from the front, standing fully erect, although note that Tyson cleverly provides it with a walking stick to indicate the difficulty that he couldn't help observing in this mode of progress! Tyson writes: "Being weak, the better to support him, I have given him a stick in his right-hand." The second plate depicts the chimp from the back, upright again, but this time holding on to a rope above its head for support! Finally, the plates of musculature and skeletal system are all portrayed in a fully upright human posture.

Front view of Tyson's pygmie from his 1699 treatise. Note how he reconstructed the animal as walking erect to enhance its human-like features. But it did not walk this way in life, so Tyson supplied a walking stick. FROM TYSON, 1699.

Back view of Tyson's chimp, this time holding on to a rope for support. FROM TYSON, 1699.

In many other passages, Tyson awards almost human attributes and emotions to his pygmy. He recalls with delight, for example, how the chimp loved to wear clothes and put them on while in bed, although he noted that it never learned not to perform nature's functions in the same place as well:

> After our pygmie was taken, and a little used to wear cloaths, it was fond enough of them; and what it could not put on himself, it would bring in his hands to some of the company to help him to put on. It would lie in a bed, place his head on the pillow, and pull the cloaths over him, as a man would do; but was so careless, and so very a brute, as to do all nature's occasions there.

Often, Tyson discussed the chimp's behavior in purely human terms: "For I heard it cry myself like a child; and he hath been often seen to kick with his feet, as children do, when either he was pleased or angered." In one passage, he even grants superiority to his chimp in matters of temperance:

> Once it was made drunk with punch, (and they are fond enough of strong liquors) but it was observed, that after that time, it would never drink above one cup, and refused the offer of more than what he found agreed with him. Thus we see instinct of nature teaches brutes temperance; and intemperance is a crime not only against the laws of morality, but of nature too.

As a second reason for exaggerating similarities between his chimp and humans, Tyson made a crucial error. He knew that his pygmy was a young animal, for extremities of the long bones were still formed in cartilage and not fully ossified, but he regarded it as nearly full grown because he mistook the complete set of milk teeth for a permanent dentition (the baby teeth of great apes do, in some respects, resemble the permanent teeth of humans). Thus, he did not realize how young an animal—a baby almost—he was dissecting. (This misidentification also enhanced his subsequent error, in a philological treatise appended to his anatomy, of attributing classical legends and more modern reports of African Pygmies to the same animal, which he regarded as just over two feet tall when fully grown.)

I have often discussed in these essays the role of neoteny (literally, holding on to youth) in human evolution (see *Ever Since Darwin* and *The Panda's Thumb*). We have evolved by slowing down the general developmental rates of primates and other mammals. Thus, human adults resemble juvenile chimps and gorillas much more closely than adult great apes. Consequently, the skeleton of a baby chimpanzee will retain many humanlike characters that an adult would lose —including a relatively large head (human babies, of course, also have relatively larger heads than human adults), a more upright mounting of the head on the spine (since the *foramen magnum,* or hole of articulation between skull and spinal column, moves back with growth), a more bulbous cranium (since the brain grows much more slowly than the body after birth), weaker brow ridges, and smaller jaws. Tyson's plate of his pygmy's skeleton, a remarkably accurate figure (I have seen photos of the original bones), shows all these humanlike features.

Tyson also noted all these features with delight in his text, but missed the coordinating theme—not that chimps are so like humans, but that he had dissected a very young animal and juvenile primates resemble adult humans in many ways, without demonstrating direct descent or relationship. He wrote, for example:

> As for the face of our pygmie, it was liker a man's than ape's and monkeys faces are: for it's forehead was larger, and more globous, and the upper and lower jaw not so long or prominent, and more spread; and it's head more than as big again as either of theirs.

Indeed, the large and humanlike brain of Tyson's chimp posed quite a problem. Tyson had already determined that

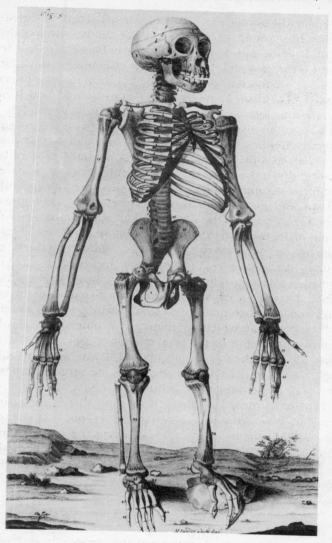

The skeleton of Tyson's chimp, again with human-like features exaggerated in position of skull on spinal column, upright posture, and subtle details of proportions. A classic example of the use of illustration to demonstrate a point (or illustrate a bias). FROM TYSON, 1699.

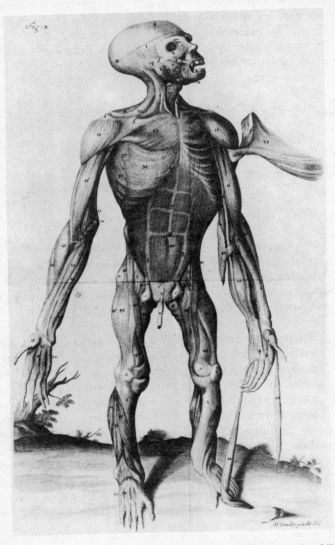

The musculature of Tyson's chimp. The bizarre "peeling back" of
exterior muscles (to show others within) was a convention of
anatomical illustration at the time. FROM TYSON, 1699.

the vocal apparatus of his pygmy was sufficiently similar to our own for speech, so why did it not talk? Perhaps a deficiency of brain prevented the expression of this most human attribute. Yet Tyson found little difference, either in basic structure or relative size, between his pygmy's brain and our own.

> One would be apt to think, that since there is so great a disparity between the soul of a man, and a brute, the organ likewise in which 'tis placed should be very different too. Yet by comparing the brain of our pygmie with that of a man, and with the greatest exactness, observing each part in both; it was very surprising to me to find so great a resemblance of the one to the other, that nothing could be more.

In a fascinating passage, displaying the seventeenth-century context of his work, Tyson simply denied that physical structure must provide a key to function. The brains are similar indeed, but humans possess some higher principle that potentiates the same matter in a different way:

> There is no reason to think, that agents do perform such and such actions, because they are found with organs proper thereunto; for then our pygmie might be really a man. The organs in animal bodies are only a regular compages of pipes and vessels, for the fluids to pass through, and are passive. What actuates them, are the humours and fluids: and animal life consists in their due and regular motion in this organical body. But those nobler faculties in the mind of man, must certainly have a higher principle; and matter organized could never produce them; for why else, where the organ is the same, should not the actions be the same too?

If the chain of being had enduring value as a heuristic prod for the exploration of missing links, and if gaps grew greater as the chain advanced, then what about the chasm even more glaring than the one that Tyson thought he had filled between ape and human—between humans and angels or other celestial beings? Tyson gave the problem a cursory comment, more political than scientific, by suggesting in his dedicatory epistle to John Sommers, Lord High Chancellor of England and President of the Royal Society (publishers of his treatise), that men of such ample learning might well plug the gap themselves!

> The animal of which I have given the anatomy, coming nearest to mankind; seems the nexus of the animal and rational, as your Lordship, and those of your High Rank and Order for knowledge and wisdom, approaching nearest to that kind of beings which is next above us; connect the visible, and invisible world.

Yet, though Tyson didn't pursue this issue, the chain's gap between human and angel became a major impetus for early speculations about a subject currently popular and, for the first time, perhaps approachable—exobiology (see essays in part 7). For the obvious solution must contend that creatures more advanced than humans, and plugging the gap between man and angel, inhabit other planets. The philosopher Immanuel Kant, for example, argued that a large and heavy planet like Jupiter must support such higher creatures. And Alexander Pope gave them explicit notice in his couplets on the chain of being from his *Essay on Man* (while praising Isaac Newton as a paragon of earthly wisdom at the same time):

> Superior beings, when of late they saw,
> A mortal man unfold all nature's law,
> Admired such wisdom in an earthly shape
> And show'd a Newton as we show an ape.

Pope only indulged in reveries framed in heroic couplets. Tyson was the man who first showed an ape with accuracy and admirable thoroughness.

18 | Bound by the Great Chain

IN *A Child's Garden of Verses,* Robert Louis Stevenson labeled the following couplet as a Happy Thought:

The world is so full of a number of things,
I'm sure we should all be as happy as kings.

Yet most of us do not rejoice when we contemplate the overwhelming diversity of nature; we are stunned by complexity and confusion. We cannot be satisfied until we have established some kind of order; we must make sense of the bewildering variety by classifying it.

Evolution is a satisfying ordering principle and we use it without hesitation today, for evolution both records the pathway of nature and allows us to classify organisms in a coherent manner. But what systems did scientists use before evolution became so popular during the nineteenth century? The "great chain of being," or even gradation of all living things, surely held pride of place among all competitors. Arthur Lovejoy, the celebrated historian of ideas who traced the lineage of this notion in his greatest work (see bibliography), called the chain of being "one of the half-dozen most potent and persistent presuppositions in Western thought. It was, in fact, until not much more than a century ago probably the most widely familiar conception of the general scheme of things, of the constitutive pattern of the universe."

In the great chain of being, each organism forms a definite link in a single sequence leading from the lowest amoeba in a drop of water to ever more complex beings, culminating in, you guessed it, our own exalted selves.

> Mark how it mounts to man's imperial race,
> from the green myriads in the peopled grass.

wrote Alexander Pope in his expostulations in heroic couplets from the *Essay on Man*.

Since we tend to confuse evolution with progress, the chain of being has often been misinterpreted as a primitive version of evolutionary theory. Although some nineteenth-century thinkers, in Lovejoy's words, "temporalized" the chain and converted it into a ladder that organisms might climb in their evolutionary advance, the original chain of being was explicitly and vehemently antievolutionary. The chain is a static ordering of unchanging, created entities—a set of creatures placed by God in fixed positions of an ascending hierarchy representing neither time nor history, but the eternal order of things. The static nature of the chain defines its ideological function: Each creature must be satisfied with its assigned place—the serf in his hovel as well as the lord in his castle—for any attempt to rise will disrupt the universe's established order. Again, Alexander Pope:

> From Nature's chain whatever link you strike,
> Tenth, or ten thousandth, breaks the chain alike.

In this essay I shall analyze the arguments presented in England's last influential defense of the chain as a static order—physician and biologist Charles White's 1799 treatise, "An account of the regular gradation in man, and in different animals and vegetables." Charles White (1728–1813), who lived and practiced in Manchester, England, was a surgeon renowned for his work in obstetrics, particularly for his insistence on absolute cleanliness during delivery. In

1795, he presented his thoughts on the chain of being to the Literary and Philosophical Society of Manchester. He published the results four years later.

For this conservative physician, the chain functioned in its usual way as an ideological support for social stability and traditional values. From the static nature of the chain itself, White inferred the necessary existence of God as a creative agent—for the only alternative would convert the chain into a temporal product of evolution, a clearly unacceptable interpretation. In the last line of his treatise, White justifies his labors by writing that "whatever tends to display the wisdom, order, and harmony of the creation, and to evince the necessity of recurring to a Deity as a first cause, must be agreeable to man." And although White expressed his opposition to slavery, and insisted that he merely wished to examine a proposition in natural history, not to cast aspersions upon any race, his conventional ranking of human groups with European whites on top and African blacks on the bottom certainly reinforced the prejudices of his comfortable Caucasian contemporaries. White insisted, speaking of himself:

> Neither is he desirous of assigning to any one a superiority over another, except that which naturally arises from superior bodily strength, mental powers, and industry, or from the consequences attendant upon living in a state of society. He only wishes to investigate the truth, and to discover what are the established laws of nature respecting his subject; apprehending, that whatever tends to elucidate the natural history of mankind, must be interesting to man.

The chain of being had always vexed biologists because, in some objective sense, it doesn't seem to describe nature very well. How can we arrange all organisms in a single, finely gradated chain when enormous gaps seem to pervade nature's system—what comes between plants and animals or invertebrates and vertebrates, for example? And how can we place into a hierarchy of perfection those creatures that seem to represent equivalent variations of a basic design, not lower or higher productions—the breeds of dogs, for example, or the persistent dilemma of human racial diversity?

In an important way the chain of being had always been a bad argument, even in its own terms and for its own time —at least if one believes that a theory about nature should record its literal appearance accurately (a criterion not always in vogue among the learned). Paradoxically, this very feature of poor harmony with nature makes the chain of being a particularly interesting subject for analysis. Good arguments don't provide nearly as much insight into human thought, for we can simply say that we have seen nature aright and have properly pursued the humble task of mapping things accurately and objectively. But bad arguments must be defended in the face of nature's opposition, a task that takes some doing. The analysis of this "doing" often provides us with insight into the ideology or thought processes of an age, if not into the modes of human reasoning itself. White's defense of the static chain is particularly forthright and unsubtle, but no different in substance from other, more sophisticated versions. It thus becomes an excellent primer for the construction of dubious arguments.

White regarded the various human races as separately created species (consistent with his antievolutionary view of gradation in the chain of being) and devoted his treatise to ordering these races as a single sequence from lower to higher. His book pursues two difficult arguments (in sequence) to reach its dubious conclusion. First, White must justify the chain of being in general, and amidst the large gaps that seem to separate plants from animals, and apes from humans. Second, he must arrange human races in a single chain, even though their variation is so multifarious that diverse criteria seem to yield different orderings. In short, how do you construct a single chain when nature seems to present abundant variation but little hierarchy?

The first part of White's treatise attempts to justify the chain as a general ordering principle for all of life. He first tackles the problem of apparent gaps between major kingdoms, plants and animals in particular. Previous advocates of the chain had generally "resolved" this dilemma by proposing fanciful arguments for intermediate forms. Thus, Charles Bonnet advocated asbestos as transitional between minerals and plants because its fibrous nature recalled the vascular systems of plants. And the freshwater hydra, a relative of corals, was widely heralded, after its discovery in 1739, as an intermediate form between plants and animals because (like plants) it seemed to lack complex internal organs and it propagated asexually by budding.

White paid traditional homage to hydras, but his main strategy for bridging the gap between plants and animals invoked an argument for similarity of anatomical design—for if he could show that plants and animals did not differ in basic design, but proceeded from the same mold with plants as less complex versions of the same fundamental plan, then a single order could be constructed. White proposed three poor arguments in attempting to establish a unity of structure between plants and animals. First, he invoked some bad analogies in claiming, for example, that since plants drop their leaves and mammals shed their hair, a fundamental similarity unites bushes and baboons. Second, he plied simple misinformation in claiming that plants have lungs for breathing. Third, he cited similarities now judged irrelevant because they are too general to support any claim for structural similarity—for example, that plants, as well as animals, are subject to disease.

To plug the largest perceived gap at the other end of the scale—that between apes and humans (although it seems smaller to us today)—White employed the same poor arguments. He did not bother to establish unity of design (even an ape's biggest detractor could not deny anatomical similarity with humans). Instead, he tried to raise the status of apes while lowering the worth of supposedly inferior people. Using bad analogies (or transferring human concepts to animal behavior), he argued that baboons assign sentinels to watch over their sleeping herds by night. In an amusing passage, and on the theme of simple misinformation, White elevated orangutans by arguing that they willingly submit to that most enlightened of contemporary medical practices—bloodletting: "When sick, these animals have been known to suffer themselves to be blooded, and even to invite the operation; and to submit to other necessary treatment, like rational creatures." Then, in a double whammy designed to raise apes and debase black humans, he portrayed the simians as both slavers and sexual abusers (ever so human, if not particularly admirable):

They have been known to carry off negro-boys, girls, and even women, with a view of making them subservient to their wants as slaves, or as objects of brutal passion: and it has been asserted by some, that women have had offspring from such connections.

Having thus established the chain as a finely nuanced sequence embracing all living things, White proceeded to the major subject of his treatise: the ranking of human races in a single order with his own group on top. For more than 100 pages, structure after structure and organ after organ, White strives mightily to arrange the races as a single sequence. The effort was an intellectual struggle involving the uncomfortable fit of recalcitrant data into a predetermined scheme; for differences among races cannot easily be linearized, no matter how strong one's a priori commitment to such an arrangement. Moreover, when we force characters into single sequences, we cannot always establish the same directions for each character—blacks may exhibit less of some admirable qualities than whites, but whites will surely rank lower for other features. How did White deal with these inconsistencies and threats to his system?

I can make most sense of White's efforts by arranging his discussions of particular features into four categories—and by noting that only one comfortably matches his preferred scheme of a single chain rising from "lower" animals to "inferior" races (African blacks at the bottom and Orientals in the middle) and finally to European whites at the summit. The first category includes admirable traits possessed in greater quantity by whites, lesser by blacks, and still less by

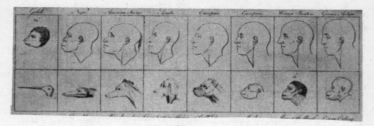

Charles White's heterogeneous version of the chain of being. Note particularly the "ascent" of human races to the abstract ideal of Greek statuary. FROM WHITE, 1799. REPRINTED FROM NATURAL HISTORY.

beasts. For example, using some dubious measures (for human races do not differ substantially in the size of their brains, as if it mattered), White argued that blacks occupied an intermediate position in a heterogeneous sequence of brain size, ranging from birds to dogs to apes and finally through "lower" human races to white Europeans (see figure of White's diversely cobbled chain of being above). But only this category among his four affirmed White's presuppositions. The other three imposed distinct and pressing problems in interpretation. White, however, was equal to the task.

The second category includes those admirable traits that, to White's embarrassment, are more abundantly distributed among black people. White dealt with this dilemma by arguing that, although the traits must be deemed valuable, beasts are even better endowed—so the sequence still runs from beast to black to white. He writes: "In these last particulars the order is changed, the European being the lowest, the African higher, and the brute creation still higher in the scale." Blacks, for example, sweat less than whites—a seeming advance in refinement (although White assures us that blacks have a stronger body odor than Caucasians). White comments:

Captains and Surgeons of Guinea ships, and the West India planters, unanimously concur in their accounts,

that negroes sweat much less than Europeans; a drop
of sweat being scarcely ever seen upon them. Simiae
sweat still less, and dogs not at all.

Similarly, black females have less copious menstruation—a
clear increment in daintiness over whites. But most apes
bleed even less or not at all. Blacks excel whites in memory,
but lower animals are the all-time champs; elephants truly
never forget. Indeed, White manages to degrade anything
admirable about blacks by attributing more of the same to
lower animals. Blacks, he claims for example, tolerate pain
better than whites. He cites a colleague who wrote:

> They bear surgical operations much better than white
> people; and what would be the cause of insupportable
> pain to a white man, a negro would almost disregard.
> I have amputated the legs of many negroes, who have
> held the upper part of the limb themselves.

But think of how many lower animals—insects in particular
—bear dismemberment without an apparent whimper.

The third category includes bestial features possessed
more strongly by whites than blacks, but even more intensely by lower animals—the most direct and evident exception to White's preferred order. Whites, for example,
have a fuller beard and more copious body hair than blacks,
while most mammals are fully covered with a dense pelage.
White wriggles out of this problem with a rhetorical device
and a claim that the noblest of animals have flowing hair,
like the copious locks of European whites!

> The fine, long, flowing hair appears to be given for
> ornament. The Universal Parent has bestowed it upon
> but few animals, and those of the noblest kind—upon
> man, the chief of the creation—upon the majestic lion,
> the king of the forest—and upon that most beautiful
> and useful domestic animal, the horse.

In the final category, blacks possess more of apparently
bestial features than whites, so all seems well—until we

realize that beasts are the least endowed of all. Black males, for example, have larger penises than whites, while black females have larger breasts—sure signs of an indecent and unbridled sexuality. (White even reports that "the Hottentot women have long flabby breasts; and that they can suckle their children upon their backs by throwing the breasts over their shoulders.") But apes have smaller penises and breasts than any group of humans. White found no adequate solution to this problem and simply made an end run around it, commenting in passing that at least black women and apes develop the largest nipples!

At this point, and after 100 pages of assiduous listing, White's argument lies in a shambles—despite all his heroic efforts to patch it up, as documented in the foregoing discussion. Therefore, following all the old adages about putting the best face upon adversity, he ends with a rhetorical flourish and with a blatant appeal to that ultimate subjectivity—aesthetic criteria. After all, don't we all know that white people are more attractive and pleasing to God and man—and that's ultimately that. Thus, in a final roulade and in a famous paragraph often quoted for its unintended humorous effect, White ends his argument with the following paean to European beauty:

> Ascending the line of gradation, we come at last to the white European; who being most removed from the brute creation, may, on that account, be considered as the most beautiful of the human race. No one will doubt his superiority of intellectual powers; and I believe it will be found that his capacity is naturally superior also to that of every other man. Where shall we find, unless in the European, that nobly arched head, containing such a quantity of brain, and supported by a hollow conical pillar, entering its center? Where the perpendicular face, the prominent nose, and round projecting chin? Where that variety of features, and fulness of expression; those long, flowing, graceful ringlets; that majestic beard, those rosy cheeks and coral lips? Where that erect posture of the body and noble gait? In what other quarter of the globe shall we

290 | THE FLAMINGO'S SMILE

find the blush that overspreads the soft features of the beautiful women of Europe, that emblem of modesty, of delicate feelings, and of sense? Where that nice expression of the amiable and softer passions in the countenance; and that general elegance of features and complexion? Where, except on the bosom of the European woman, two such plump and snowy white hemispheres, tipt with vermillion?

I don't mean to diminish the posthumous humor of this passage—"snowy white hemispheres tipt with vermillion" as the ultimate mark of human perfection, indeed! White's flowery style may render him more subject to ridicule than most of his contemporaries, but his argument is no worse or different from many of theirs. He was merely expressing a common opinion of his time in admittedly overblown rhetoric. The static chain of being, as Lovejoy argues, had formed a cornerstone of Western interpretations of nature for centuries, despite its evident difficulties in application to a recalcitrant world full of gaps and copious variation not easily ordered into single sequences.

So have a good chuckle at the appropriate parts, but then ponder the larger and serious issue for a moment. Evolution drove the static chain of being into obsolescence—therefore, we may easily, in retrospect, identify its evident flaws and analyze the falseness and inconsistency of argument used to defend it. But how many of our own cherished beliefs, the ones that we never doubt because we think that they map nature in an obvious way, will seem centuries hence just as foolish and ideologically bound as the static chain of being? Should we not examine the logic and verisimilitude of our own deepest convictions? At least we may avoid the ridicule of future generations by steering clear of sexual anatomy and leaving to the great biblical poets of the Song of Songs any metaphorical description of the human breast.

19 | The Hottentot Venus

I HAD A LITTLE FRIEND in nursery school. I don't even remember her name. But I do recall some secret advice that I offered her one day at the playground. I told her that the enormous surrounding creatures known as adults always looked up when they walked, and that we little folk would therefore find all manner of valuable things on the ground if only we kept our gazes down. Were my paleontological predispositions already in evidence?

Carl Sagan and I both grew up in New York, both interested in biology and astronomy. Since Carl is tall and chose astronomy, while I'm short and chose paleontology, I always figured that he'd be looking up (as he did with some regularity in hosting his TV series *Cosmos*), while I'd be sticking to my old but good advice and staring at the ground. But I one-upped him (literally) last month in Paris.

A few years back, Yves Coppens, professor at the Musée de l'Homme in Paris, took Carl on a tour of the museum's innards. There, on a shelf in storage, they found the brain of Paul Broca floating in Formalin in a bell jar. Carl wrote a fine essay about this visit, the title piece of his book *Broca's Brain.* A few months ago, Yves took me on a similar tour. I held the skull of Descartes and of our mutual ancestor, the old man of Cro-Magnon. I also found Broca's brain, resting on its shelf and surrounded by other bell jars holding the brains of his illustrious scientific contemporaries—all white and all male. Yet I found the most interesting items on the shelf just above. Perhaps Carl never looked up.

This area of the museum's "back wards" holds Broca's collection of anatomical parts, including his own generous and posthumous contribution. Broca, a great medical anatomist and anthropologist, embodied the great nineteenth-century faith in quantification as a key to objective science. If he could collect enough human parts from enough human races, the resultant measurements would surely define the great scale of human progress, from chimp to Caucasian. Broca was not more virulently racist than his scientific contemporaries (nearly all successful white males, of course); he was simply more assiduous in accumulating irrelevant data, selectively presented to support an a priori viewpoint.

These shelves contain a ghoulish potpourri: severed heads from New Caledonia; an illustration of foot binding as practiced upon Chinese women—yes, a bound foot and lower leg, severed between knee and ankle. And, on a shelf just above the brains, I saw a little exhibit that provided an immediate and chilling insight into nineteenth-century *mentalité* and the history of racism: in three smaller jars, I saw the dissected genitalia of three Third-World women. I found no brains of women, and neither Broca's penis nor any male genitalia grace the collection.

The three jars are labeled *une négresse, une péruvienne,* and *la Vénus Hottentotte,* or the Hottentot Venus. Georges Cuvier himself, France's greatest anatomist, had dissected the Hottentot Venus upon her death in Paris late in 1815. He went right to the genitalia for a particular and interesting reason, to which I will return after recounting the tale of this unfortunate woman.

In an age before television and movies made virtually nothing on earth exotic, and when anthropological theory assessed as subhuman both malformed Caucasians and the normal representatives of other races, the exhibition of unusual humans became a profitable business both in upper-class salons and in street-side stalls (see Richard D. Altick's *The Shows of London,* in the bibliography, or the book, stage, and screen treatments of the "Elephant Man").

Supposed savages from faraway lands were a mainstay of these exhibitions, and the Hottentot Venus surpassed them all in renown. (The Hottentots and Bushmen are closely related, small-statured people of southern Africa. Traditional Bushmen, when first encountered by Europeans, were hunter-gatherers, while Hottentots were pastoralists who raised cattle. Anthropologists now tend to forgo these European, somewhat derogatory terms and to designate both groups collectively as the Khoi-San peoples, a composite word constructed from each group's own name for itself.) The Hottentot Venus was a servant of Dutch farmers near Capetown, and we do not know her actual group membership. She had a name, though her exploiters never used it. She was baptized Saartjie Baartman (Saartjie, or "little Sarah" in Afrikaans, is pronounced Sar-key).

Hendrick Cezar, brother of Saartjie's "employer," suggested a trip to England for exhibition and promised to make Saartjie a wealthy woman thereby. Lord Caledon, governor of the Cape, granted permission for the trip but later regretted his decision when he understood its purposes more fully. (Saartjie's exhibition aroused much debate and she always had supporters, disgusted with the display of humans as animals; the show went on, but not to universal approbation.) She arrived in London in 1810 and immediately went on exhibition in Piccadilly, where she caused a sensation, for reasons soon to be discussed. A member of the African Association, a benevolent society that petitioned for her "release," described the show. He first encountered Saartjie in a cage on a platform raised a few feet above the floor:

> On being ordered by her keeper, she came out. . . .
> The Hottentot was produced like a wild beast, and
> ordered to move backwards and forwards and come
> out and go into her cage, more like a bear in a chain
> than a human being.

Yet Saartjie, interrogated in Dutch before a court, insisted that she was not under restraint and understood perfectly

294 | THE FLAMINGO'S SMILE

well that she had been guaranteed half the profits. The show went on.

After a long tour of the English provinces, Saartjie went to Paris where an animal trainer exhibited her for fifteen months, causing as great a sensation as in England. Cuvier and all the great naturalists of France visited her and she posed in the nude for scientific paintings at the Jardin du Roi. But she died of an inflammatory ailment on December 29, 1815, and ended up on Cuvier's dissecting table, rather than wealthy in Capetown.

Why, in an age deluged with human exhibitions, was Saartjie such a sensation? We may offer two answers, each troubling and each associated with one of her official titles —Hottentot and Venus.

On the racist ladder of human progress, Bushmen and Hottentots vied with Australian aborigines for the lowest rung, just above chimps and orangs. (Some scholars have argued that the earliest designation applied by seventeenth-century Dutch settlers—*Bosmanneken,* or "Bushman"—was a literal translation of a Malay word well known to them— *Orang Outan,* or "man of the forest.") In this system, Saartjie exerted a grim fascination, not as a missing link in a later evolutionary sense, but as a creature who straddled that dreaded boundary between human and animal and thereby taught us something about a self still present, although submerged, in "higher" creatures (see essays 17 and 18).

Contemporary commentators emphasized both the simian appearance and the brutal habits of Bushmen and Hottentots. In 1839, the leading American anthropologist S.G. Morton labeled Hottentots as "the nearest approximation to the lower animals. . . . Their complexion is a yellowish brown, compared by travellers to the peculiar hue of Europeans in the last stage of jaundice. . . . The women are represented as even more repulsive in appearance than the men." Mathias Guenther (see bibliography) cites an 1847 newspaper account of a Bushman family displayed at the Egyptian Hall in London:

In appearance they are little above the monkey tribe. They are continually crouching, warming themselves

THE HOTTENTOT VENUS | 295

by the fire, chatting or growling. . . . They are sullen, silent and savage—mere animals in propensity, and worse than animals in appearance.

And the jaundiced account of a failed missionary in 1804:

The Bushmen will kill their children without remorse, on various occasions; as when they are ill shaped, or when they are in want of food, or when obliged to flee from the farmers or others; in which case they will strangle them, smother them, cast them away in the desert or bury them alive. There are instances of parents throwing their tender offspring to the hungry lion, who stands roaring before their cavern, refusing to depart before some peace offering be made to him.

Guenther reports that this equation of Bushman and animal became so ingrained that one party of Dutch settlers, out on a hunting expedition, shot and ate a Bushman, assuming that he was the African equivalent of the Malay orang.

Cuvier's monograph of Saartjie's dissection, published in the *Mémoires du Muséum d'Histoire Naturelle* for 1817, followed this traditional view. After discussing and dismissing various ill-founded legends, Cuvier promised to present only "positive facts"—including this description of a Bushman's life:

Since they are unable to engage in agriculture, or even in a pastoral life, they subsist entirely on hunting and pilfering. They live in caves and cover themselves only with the skins of animals they have killed. Their only industry involves the poisoning of their arrows and the manufacture of nets for fishing.

His description of Saartjie herself emphasizes all points of superficial similarity with any ape or monkey. (I need hardly mention that since people vary so much, each group must be closer than others to some feature of some other primate, without implying anything about genealogy or ap-

titude.) Cuvier, for example, discusses the flatness of Saartjie's nasal bones: "In this respect, I have never seen a
human head more similar to that of monkeys." He emphasizes various proportions of the femur (upper leg bone) as
embodying "characters of animality." He speaks of Saartjie's small skull (no surprise for a woman four and a half
feet tall), and relegates her to stupidity according to "that
cruel law, which seems to have condemned to an eternal
inferiority those races with small and compressed skulls."
He even abstracted a set of supposedly simian responses
from her behavior: "Her movements had something
brusque and capricious about them, which recall those of
monkeys. She had, above all, a way of pouting her lips, in
the same manner as we have observed in orang utans."

Yet a careful reading of the entire monograph belies
these interpretations, since Cuvier states again and again
(although he explicitly draws neither moral nor message)
that Saartjie was an intelligent woman with general proportions that would not lead connoisseurs to frown. He mentions, in an offhand sort of way, that Saartjie possessed an
excellent memory, spoke Dutch rather well, had some command of English, and was learning a bit of French when she
died. (Not bad for a caged brute; I only wish that more
Americans could do one-third so well in their command of
languages.) He admitted that her shoulders, back, and chest
"had grace"; and with the gentilesse of his own race, spoke
of *sa main charmante* ("her charming hand").

Yet Saartjie's hold over well-bred Europe did not arise
from her racial status alone. She was not simply the Hottentot or the Hottentot woman, but the Hottentot *Venus*.
Under all official words lay the great and largely unsaid
reason for her popularity. Khoi-San women do exaggerate
two features of their sexual anatomy (or at least of body
parts that excite sexual feelings in most men). The Hottentot Venus won her fame as a sexual object, and her combination of supposed bestiality and lascivious fascination
focused the attention of men who could thus obtain both
vicarious pleasure and a smug reassurance of superiority.

Primarily—for, as they say, you can't miss it—Saartjie

was, in Altick's words, "steatopygous to a fault." Khoi-San women accumulate large amounts of fat in their buttocks, a condition called steatopygia. The buttocks protrude far back, often coming to a point at their upper extremity and sloping down toward the genitalia. Saartjie was especially well endowed, the probable cause of Cezar's decision to convert her from servant to siren. Saartjie covered her genitalia during exhibitions, but her rear end *was* the show, and she submitted to endless gaze and poke for five long years. Since European women did not wear bustles at the time, but indicated by their clothing only what nature had provided, Saartjie seemed all the more incredible.

Cuvier well understood the mixed bestial and sexual nature of Saartjie's fascination when he wrote that "everyone was able to see her during her eighteen-month stay in our capital, and to verify the enormous protrusion of her buttocks and the brutal appearance of her face." In his dissection, Cuvier focused on an unsolved mystery surrounding each of her unusual features. Europeans had long wondered whether the large buttocks were fatty, muscular, or perhaps even supported by a previously unknown bone. The problem had already been solved—in favor of fat—by external observation, the primary reason for her disrobing before scientists at the Jardin du Roi. Still, Cuvier dissected her buttocks and reported:

> We could verify that the protuberance of her buttocks had nothing muscular about it, but arose from a [fatty] mass of a trembling and elastic consistency, situated immediately under her skin. It vibrated with all movements that the woman made.

But Saartjie's second peculiarity provided even greater wonder and speculation among scientists; and Saartjie heightened the intrigue by keeping this feature scrupulously hidden, even refusing a display at the Jardin. Only after her death could the curiosity of science be slaked.

Reports had circulated for two centuries of a wondrous structure attached directly to the female genitalia of Khoi-

San women and covering their private parts with a veil of skin, the so-called *sinus pudoris,* or "curtain of shame." (If I may be permitted a short excursion into the realm of scholarly minutiae—the footnotes of more conventional academic publication—I would like to correct a standard mistranslation of Linnaeus, one that I have made myself. In his original description of *Homo sapiens,* Linnaeus provided a most unflattering account of African blacks, including the line: *feminae sinus pudoris.* This phrase has usually been translated, "women are without shame"—a slur quite consistent with Linnaeus's general description. In Latin, "without shame" should be *sine pudore,* not *sinus pudoris.* But eighteenth-century scientific Latin was written so indifferently that misspellings and wrong cases are no bar to actual intent, and the reading "without shame" has held. But Linnaeus was only stating that African women have a genital flap, or *sinus pudoris.* He was also wrong, because only the Khoi-San and a few related peoples develop this feature.)

The nature of the *sinus pudoris* had generated a lively debate, with partisans on both sides claiming eyewitness support. One party held that the *sinus* was simply an enlarged part of the ordinary genitalia; others called it a novel structure found in no other race. Some even described the so-called "Hottentot apron" as a large fold of skin hanging down from the lower abdomen itself.

Cuvier was determined to resolve this old argument; the status of Saartjie's *sinus pudoris* would be the primary goal of his dissection. Cuvier began his monograph by noting: "There is nothing more famous in natural history than the *tablier* (the French rendering of *sinus pudoris*) of Hottentots, and, at the same time, no feature has been the object of so many arguments." Cuvier resolved the debate with his usual elegance: the *labia minora,* or inner lips, of the ordinary female genitalia are greatly enlarged in Khoi-San women, and may hang down three or four inches below the vagina when women stand, thus giving the impression of a separate and enveloping curtain of skin. Cuvier preserved his skillful dissection of Saartjie's genitalia and wrote with a flourish: "I have the honor to present to the Academy the genital organs of this woman prepared in a manner that leaves no doubt about the nature of her *tablier*." And Cuvier's gift still rests in its jar, forgotten on a shelf at the Musée de l'Homme—right above Broca's brain.

Yet while Cuvier correctly identified the nature of Saartjie's *tablier*, he fell into an interesting error, arising from the same false association that had inspired public fascination with Saartjie—sexuality with animality. Since Cuvier regarded Hottentots as the most bestial of people, and since they had a large *tablier*, he assumed that the *tablier* of other Africans must become progressively smaller as the darkness of southern Africa ceded to the light of Egypt. (In the last part of his monograph, Cuvier argues that the ancient Egyptians must have been fully Caucasian; who else could have built the pyramids?)

Cuvier knew that female circumcision was widely practiced in Ethiopia. He assumed that the *tablier* must be at least half-sized among these people of intermediate hue and geography; and he further conjectured that Ethiopians excised the *tablier* to improve sexual access, not that circumcision represented a custom sustained by power and imposed upon girls with genitalia not noticeably different from those of European women. "The negresses of Abyssinia," he wrote, "are inconvenienced to the point of being obliged to destroy these parts by knife and cauterization" (*par le fer et par le feu,* as he wrote in more euphonious French).

Cuvier also told an interesting tale, requiring no comment in repetition:

> The Portuguese Jesuits, who converted the King of Abyssinia and part of his people during the 16th century, felt that they were obliged to proscribe this practice [of female circumcision] since they thought that it was a holdover from the ancient Judaism of that nation. But it happened that Catholic girls could no longer find husbands, because the men could not reconcile themselves to such a disgusting deformity. The College of Propaganda sent a surgeon to verify the fact and, on his report, the reestablishment of the ancient custom was authorized by the Pope.

I needn't burden you with any detailed refutation of the general arguments that made the Hottentot Venus such a sensation. I do, however, find it amusing that she and her people are, by modern convictions, so singularly and especially unsuited for the role she was forced to play.

If earlier scientists cast the Khoi-San peoples as approximations to the lower primates, they now rank among the heroes of modern social movements. Their languages, with complex clicks, were once dismissed as a guttural farrago of beastly sounds. They are now widely admired for their complexity and subtle expression. Cuvier had stigmatized the hunter-gatherer life styles of the traditional San (Bushmen) as the ultimate degradation of a people too stupid and indolent to farm or raise cattle. The same people have become models of righteousness to modern ecoactivists for their understanding, nonexploitive, and balanced approach to natural resources. Of course, as Guenther argues in his article on the Bushman's changing image, our modern accolades may also be unrealistic. Still, if people must be exploited rather than understood, attributions of kindness and heroism sure beat accusations of animality.

Furthermore, while Cuvier's contemporaries sought physical signs of bestiality in Khoi-San anatomy, anthropologists now identify these people as perhaps the most paedomorphic of human groups. Humans have evolved by a general retardation (or slowing down) of developmental rates, leaving our adult bodies quite similar in many respects to the juvenile, but not to the adult, form of our primate ancestors—an evolutionary result called paedomorphosis, or "child shaping." On this criterion, the greater the extent of paedomorphosis, the further away from a simian past (although minor differences among human races do not translate into variations in mental or moral worth). Although Cuvier searched hard to find signs of animality in Saartjie's lip movements or in the form of her leg bone, her people are, in general, perhaps the least simian of all humans.

Finally, the major rationale for Saartjie's popularity rested on a false premise. She fascinated Europeans because she had big buttocks and genitalia and because she supposedly belonged to the most backward of human groups. Everything fit together for Cuvier's contemporaries. Advanced humans (read modern Europeans) are refined, modest, and sexually restrained (not to mention hypocritical for advancing such a claim). Animals are overtly and actively sexual, and so betray their primitive character. Thus, Saartjie's exaggerated sexual organs record her animality. But the argument is, as our English friends say (and quite literally in this case), "arse about face." Humans are the most sexually active of primates, and humans have the largest sexual organs of our order. If we must pursue this dubious line of argument, a person with larger than average endowment is, if anything, more human.

On all accounts—mode of life, physical appearance, and sexual anatomy—London and Paris should have stood in a giant cage while Saartjie watched. Still, Saartjie gained her posthumous triumph. Broca inherited not only Cuvier's preparation of Saartjie's *tablier,* but her skeleton as well. In 1862, he thought he had found a criterion for arranging human races by physical merit. He measured the ratio of radius (lower arm bone) to humerus (upper arm bone), reasoning that higher ratios indicate longer forearms—a traditional feature of apes. He began to hope that objective measurement had confirmed his foregone conclusion when blacks averaged .794 and whites .739. But Saartjie's skeleton yielded .703 and Broca promptly abandoned his criterion. Had not Cuvier praised the arm of the Hottentot Venus?

Saartjie continues her mastery of Mr. Broca today. His brain decomposes in a leaky jar. Her *tablier* stands above, while her well-prepared skeleton gazes up from below. Death, as the good book says, is swallowed up in victory.

302 | THE FLAMINGO'S SMILE

Postscript

Since biological determinism won its prestige in spurious claims to objectivity via quantification (see my book, *The Mismeasure of Man*), and since Saartjie Baartman owed her oppression to this sociopolitical doctrine masquerading as science, I was amused to find that Francis Galton himself, the chief apostle of quantification (and hereditarianism), once used an ingenious technique to measure the extent of steatopygia on a Khoi-San woman. Galton, Darwin's brilliant and eccentric cousin, believed that he could put anything into numbers. He once tried to quantify the geographic distribution of female beauty by the following dubious method (as described in his autobiography, *Memories of My Life,* 1909, pp. 315–316):

> Whenever I have occasion to classify the persons I meet into three classes, "good, medium, bad," I use a needle mounted as a pricker, wherewith to prick holes, unseen, in a piece of paper, torn rudely into a cross with a long leg. I use its upper end for "good," the cross arm for "medium," the lower end for "bad." The prick holes keep distinct, and are easily read off at leisure. The object, place, and date are written on the paper. I used this plan for my beauty data, classifying the girls I passed in streets or elsewhere as attractive, indifferent, or repellent. Of course this was a purely individual estimate, but it was consistent, judging from the conformity of different attempts in the same population. I found London to rank highest for beauty; Aberdeen lowest.

His discreet method for steatopygia was, in my view, even more clever (and probably a good deal more accurate if all those high school trig proofs really work). In his *Narration of an Explorer in Tropical South Africa,* he writes (my thanks to Raymond B. Huey of the University of Washington for sending this passage to me):

> The sub-interpreter was married to a charming person, not only a Hottentot in figure, but in that respect a Venus among Hottentots. I was perfectly aghast at her development, and made inquiries upon that delicate point as far as I dared among my missionary friends . . . I profess to be a scientific man, and was exceedingly anxious to obtain accurate measurements of her shape; but there was a difficulty in doing this. I did not know a word of Hottentot, and could never therefore have explained to the lady what the object of my foot-rule could be; and I really dared not ask my worthy missionary host to interpret for me. I therefore felt in a dilemma as I gazed at her form, that gift of bounteous nature to this favoured race, which no mantua-maker, with all her crinoline and stuffing, can do otherwise than humbly imitate. The object of my admiration stood under a tree, and was turning herself about to all points of the compass, as ladies who wish to be admired usually do. Of a sudden my eye fell upon my sextant; the bright thought struck me, and I took a series of observations upon her figure in every direction, up and down, crossways, diagonally, and so forth, and I registered them carefully upon an outline drawing for fear of any mistake; this being done, I boldly pulled out my measuring-tape, and measured the distance from where I was to the place she stood, and having thus obtained both base and angles, I worked out the results by trigonometry and logarithms.

Saartjie Baartman herself continues to fascinate us across the ages; her exploitation has never really ended. In an antiquarian bookstore in Johannesberg (see essay 12), I found and bought the following remarkable print (I still cannot view it without a shudder despite its intended

A satiric French print of 1812 commenting on English fascination with the Hottentot Venus. The soldier behind her examines her steatopygia, while the lady in front pretends to tie her shoelace in order to get a peek at Saartjie's tablier.

humor, and I reproduce it here as a comment upon history and current reality that we dare not ignore). The print is a satirical French commentary (published in Paris in 1812) on English fascination with Saartjie's display. It is titled: *Les curieux en extase, ou les cordons de souliers* (The curious in ecstasy, or the shoelaces). Spectators concentrate entirely upon sexual features of the Hottentot *Venus*. One military gentleman observes her steatopygia from behind and comments, *"Oh! godem quel rosbif."* The second man in uniform and the elegantly attired lady are both trying to sneak a peak at Saartjie's tablier. (This is the subtle point that an uninformed observer would miss. Saartjie displayed her buttocks but, following the customs of her people, would never uncover her tablier). The man exclaims "how odd nature is," while the woman, hoping to get a better look from below, crouches under pretense of tying her shoes (hence the title). Meanwhile, the dog reminds us that we are all the same biological object under our various attires.

To bring the exploitation up to date, W.B. Deatrick sent me the cover of the French magazine *Photo* for May, 1982. It shows, naked, a woman who calls herself "Carolina, la Vénus hottentote de Saint-Domingue." She holds an uncorked champagne bottle in front. The fizz flies up, over her head, through the letter O of the magazine's title, down behind her back and directly into the glass, which rests, as she crouches (to mimic Saartjie's endowment), upon her outstretched buttocks.

20 | Carrie Buck's Daughter

THE LORD REALLY put it on the line in his preface to that prototype of all prescription, the Ten Commandments:

> . . . for I, the Lord thy God, am a jealous God, visiting the iniquity of the fathers upon the children unto the third and fourth generation of them that hate me (Exod. 20:5).

The terror of this statement lies in its patent unfairness—its promise to punish guiltless offspring for the misdeeds of their distant forebears.

A different form of guilt by genealogical association attempts to remove this stigma of injustice by denying a cherished premise of Western thought—human free will. If offspring are tainted not simply by the deeds of their parents but by a material form of evil transferred directly by biological inheritance, then "the iniquity of the fathers" becomes a signal or warning for probable misbehavior of their sons. Thus Plato, while denying that children should suffer directly for the crimes of their parents, nonetheless defended the banishment of a personally guiltless man whose father, grandfather, and great-grandfather had all been condemned to death.

It is, perhaps, merely coincidental that both Jehovah and Plato chose three generations as their criterion for estab-

306

lishing different forms of guilt by association. Yet we maintain a strong folk, or vernacular, tradition for viewing triple occurrences as minimal evidence of regularity. Bad things, we are told, come in threes. Two may represent an accidental association; three is a pattern. Perhaps, then, we should not wonder that our own century's most famous pronouncement of blood guilt employed the same criterion—Oliver Wendell Holmes's defense of compulsory sterilization in Virginia (Supreme Court decision of 1927 in *Buck* v. *Bell*): "three generations of imbeciles are enough."

Restrictions upon immigration, with national quotas set to discriminate against those deemed mentally unfit by early versions of IQ testing, marked the greatest triumph of the American eugenics movement—the flawed hereditarian doctrine, so popular earlier in our century and by no means extinct today (see following essay), that attempted to "improve" our human stock by preventing the propagation of those deemed biologically unfit and encouraging procreation among the supposedly worthy. But the movement to enact and enforce laws for compulsory "eugenic" sterilization had an impact and success scarcely less pronounced. If we could debar the shiftless and the stupid from our shores, we might also prevent the propagation of those similarly afflicted but already here.

The movement for compulsory sterilization began in earnest during the 1890s, abetted by two major factors—the rise of eugenics as an influential political movement and the perfection of safe and simple operations (vasectomy for men and salpingectomy, the cutting and tying of Fallopian tubes, for women) to replace castration and other socially unacceptable forms of mutilation. Indiana passed the first sterilization act based on eugenic principles in 1907 (a few states had previously mandated castration as a punitive measure for certain sexual crimes, although such laws were rarely enforced and usually overturned by judicial review). Like so many others to follow, it provided for sterilization of afflicted people residing in the state's "care," either as inmates of mental hospitals and homes for the feeble-minded or as inhabitants of prisons. Sterilization could be

Official Virginia hospital form for sexual sterilization.

imposed upon those judged insane, idiotic, imbecilic, or moronic, and upon convicted rapists or criminals when recommended by a board of experts.

By the 1930s, more than thirty states had passed similar laws, often with an expanded list of so-called hereditary defects, including alcoholism and drug addiction in some states, and even blindness and deafness in others. These laws were continually challenged and rarely enforced in most states; only California and Virginia applied them zealously. By January 1935, some 20,000 forced "eugenic" sterilizations had been performed in the United States, nearly half in California.

No organization crusaded more vociferously and successfully for these laws than the Eugenics Record Office, the semiofficial arm and repository of data for the eugenics movement in America. Harry Laughlin, superintendent of the Eugenics Record Office, dedicated most of his career to a tireless campaign of writing and lobbying for eugenic sterilization. He hoped, thereby, to eliminate in two generations the genes of what he called the "submerged tenth"—"the most worthless one-tenth of our present population." He proposed a "model sterilization law" in 1922, designed

> to prevent the procreation of persons socially inadequate from defective inheritance, by authorizing and providing for eugenical sterilization of certain potential parents carrying degenerate hereditary qualities.

This model bill became the prototype for most laws passed in America, although few states cast their net as widely as Laughlin advised. (Laughlin's categories encompassed "blind, including those with seriously impaired vision; deaf, including those with seriously impaired hearing; and dependent, including orphans, ne'er-do-wells, the homeless, tramps, and paupers.") Laughlin's suggestions were better heeded in Nazi Germany, where his model act inspired the infamous and stringently enforced *Erbgesundheitsrecht*, leading by the eve of World War II to the sterilization of some 375,000 people, most for "congenital feeblemindedness," but including nearly 4,000 for blindness and deafness.

The campaign for forced eugenic sterilization in America reached its climax and height of respectability in 1927, when the Supreme Court, by an 8–1 vote, upheld the Virginia sterilization bill in *Buck* v. *Bell.* Oliver Wendell Holmes, then in his mid-eighties and the most celebrated jurist in America, wrote the majority opinion with his customary verve and power of style. It included the notorious paragraph, with its chilling tag line, cited ever since as the quintessential statement of eugenic principles. Remembering with pride his own distant experiences as an infantryman in the Civil War, Holmes wrote:

> We have seen more than once that the public welfare may call upon the best citizens for their lives. It would be strange if it could not call upon those who already sap the strength of the state for these lesser sacrifices. . . . It is better for all the world, if instead of waiting to execute degenerate offspring for crime, or to let them starve for their imbecility, society can prevent those who are manifestly unfit from continuing their kind. The principle that sustains compulsory vaccination is broad enough to cover cutting the Fallopian tubes. Three generations of imbeciles are enough.

Who, then, were the famous "three generations of imbeciles," and why should they still compel our interest?

When the state of Virginia passed its compulsory sterilization law in 1924, Carrie Buck, an eighteen-year-old white woman, lived as an involuntary resident at the State Colony for Epileptics and Feeble-Minded. As the first person selected for sterilization under the new act, Carrie Buck became the focus for a constitutional challenge launched, in part, by conservative Virginia Christians who held, according to eugenical "modernists," antiquated views about individual preferences and "benevolent" state power. (Simplistic political labels do not apply in this case, and rarely in general for that matter. We usually regard eugenics as a conservative movement and its most vocal critics as members of the left. This alignment has generally held in our own decade. But eugenics, touted in its day as the latest in scientific modernism, attracted many liberals and numbered among its most vociferous critics groups often labeled as reactionary and antiscientific. If any political lesson emerges from these shifting allegiances, we might consider the true inalienability of certain human rights.)

But why was Carrie Buck in the State Colony and why was she selected? Oliver Wendell Holmes upheld her choice as judicious in the opening lines of his 1927 opinion:

> Carrie Buck is a feeble-minded white woman who was committed to the State Colony. . . . She is the daughter of a feeble-minded mother in the same institution, and the mother of an illegitimate feeble-minded child.

In short, inheritance stood as the crucial issue (indeed as the driving force behind all eugenics). For if measured mental deficiency arose from malnourishment, either of body or mind, and not from tainted genes, then how could sterilization be justified? If decent food, upbringing, medical care, and education might make a worthy citizen of Carrie Buck's daughter, how could the State of Virginia justify the severing of Carrie's Fallopian tubes against her will? (Some forms of mental deficiency are passed by inheritance in family lines, but most are not—a scarcely surprising conclusion when we consider the thousand shocks that beset us all during our lives, from abnormalities in embryonic growth to traumas of birth, malnourishment, rejection, and poverty. In any case, no fair-minded person today would credit Laughlin's social criteria for the identification of hereditary deficiency—ne'er-do-wells, the homeless, tramps, and paupers—although we shall soon see that Carrie Buck was committed on these grounds.)

When Carrie Buck's case emerged as the crucial test of Virginia's law, the chief honchos of eugenics understood that the time had come to put up or shut up on the crucial issue of inheritance. Thus, the Eugenics Record Office sent Arthur H. Estabrook, their crack fieldworker, to Virginia for a "scientific" study of the case. Harry Laughlin himself provided a deposition, and his brief for inheritance was presented at the local trial that affirmed Virginia's law and later worked its way to the Supreme Court as *Buck* v. *Bell*.

Laughlin made two major points to the court. First, that Carrie Buck and her mother, Emma Buck, were feeble-minded by the Stanford-Binet test of IQ, then in its own infancy. Carrie scored a mental age of nine years, Emma of seven years and eleven months. (These figures ranked them technically as "imbeciles" by definitions of the day, hence Holmes's later choice of words—though his infamous line is often misquoted as "three generations of idiots." Imbeciles displayed a mental age of six to nine years; idiots performed worse, morons better, to round out the old nomenclature of mental deficiency.) Second, that most feeble-mindedness resides ineluctably in the genes, and that Carrie Buck surely belonged with this majority. Laughlin reported:

> Generally feeble-mindedness is caused by the inheritance of degenerate qualities; but sometimes it might be caused by environmental factors which are not hereditary. In the case given, the evidence points strongly toward the feeble-mindedness and moral delinquency of Carrie Buck being due, primarily, to inheritance and not to environment.

Carrie Buck's daughter was then, and has always been, the pivotal figure of this painful case. I noted in beginning this essay that we tend (often at our peril) to regard two as potential accident and three as an established pattern. The supposed imbecility of Emma and Carrie might have been an unfortunate coincidence, but the diagnosis of similar deficiency for Vivian Buck (made by a social worker, as we shall see, when Vivian was but six months old) tipped the balance in Laughlin's favor and led Holmes to declare the Buck lineage inherently corrupt by deficient heredity. Vivian sealed the pattern—*three* generations of imbeciles are enough. Besides, had Carrie not given illegitimate birth to Vivian, the issue (in both senses) would never have emerged.

Oliver Wendell Holmes viewed his work with pride. The man so renowned for his principle of judicial restraint, who had proclaimed that freedom must not be curtailed without "clear and present danger"—without the equivalent of falsely yelling "fire" in a crowded theater—wrote of his judgment in *Buck* v. *Bell:* "I felt that I was getting near the first principle of real reform."

And so *Buck* v. *Bell* remained for fifty years, a footnote to a moment of American history perhaps best forgotten. Then, in 1980, it reemerged to prick our collective conscience, when Dr. K. Ray Nelson, then director of the Lynchburg Hospital where Carrie Buck had been sterilized, researched the records of his institution and discovered that more than 4,000 sterilizations had been performed, the last as late as 1972. He also found Carrie Buck, alive and well near Charlottesville, and her sister Doris, covertly sterilized under the same law (she was told that her operation was for appendicitis), and now, with fierce dignity, dejected and bitter because she had wanted a child more than anything else in her life and had finally, in her old age, learned why she had never conceived.

As scholars and reporters visited Carrie Buck and her sister, what a few experts had known all along became abundantly clear to everyone. Carrie Buck was a woman of obviously normal intelligence. For example, Paul A. Lombardo of the School of Law at the University of Virginia, and a leading scholar of *Buck* v. *Bell,* wrote in a letter to me:

> As for Carrie, when I met her she was reading newspapers daily and joining a more literate friend to assist at regular bouts with the crossword puzzles. She was not a sophisticated woman, and lacked social graces, but mental health professionals who examined her in later life confirmed my impressions that she was neither mentally ill nor retarded.

On what evidence, then, was Carrie Buck consigned to the State Colony for Epileptics and Feeble-Minded on January 23, 1924? I have seen the text of her commitment hearing; it is, to say the least, cursory and contradictory. Beyond the bald and undocumented say-so of her foster parents, and her own brief appearance before a commission of two doctors and a justice of the peace, no evidence was presented. Even the crude and early Stanford-Binet test, so fatally flawed as a measure of innate worth (see my book *The Mismeasure of Man*, although the evidence of Carrie's own case suffices) but at least clothed with the aura of quantitative respectability, had not yet been applied.

When we understand why Carrie Buck was committed in January 1924, we can finally comprehend the hidden meaning of her case and its message for us today. The silent key, again as from the first, is her daughter Vivian, born on March 28, 1924, and then but an evident bump on her belly. Carrie Buck was one of several illegitimate children borne by her mother, Emma. She grew up with foster parents, J. T. and Alice Dobbs, and continued to live with them as an adult, helping out with chores around the house. She was raped by a relative of her foster parents, then blamed for the resulting pregnancy. Almost surely, she was (as they used to say) committed to hide her shame (and her rapist's identity), not because enlightened science had just discovered her true mental status. In short, she was sent away to have her baby. Her case never was about mental deficiency; Carrie Buck was persecuted for supposed sexual immorality and social deviance. The annals of her trial and hearing reek with the contempt of the well-off and well-bred for poor people of "loose morals." Who really cared whether Vivian was a baby of normal intelligence; she was the illegitimate child of an illegitimate woman. Two generations of bastards are enough. Harry Laughlin began his "family history" of the Bucks by writing: "These people belong to the shiftless, ignorant and worthless class of anti-social whites of the South."

We know little of Emma Buck and her life, but we have no more reason to suspect her than her daughter Carrie of true mental deficiency. Their supposed deviance was social and sexual; the charge of imbecility was a cover-up, Mr. Justice Holmes notwithstanding.

We come then to the crux of the case, Carrie's daughter,

Vivian. What evidence was ever adduced for her mental deficiency? This and only this: At the original trial in late 1924, when Vivian Buck was seven months old, a Miss Wilhelm, social worker for the Red Cross, appeared before the court. She began by stating honestly the true reason for Carrie Buck's commitment:

> Mr. Dobbs, who had charge of the girl, had taken her when a small child, had reported to Miss Duke [the temporary secretary of Public Welfare for Albemarle County] that the girl was pregnant and that he wanted to have her committed somewhere—to have her sent to some institution.

Miss Wilhelm then rendered her judgment of Vivian Buck by comparing her with the normal granddaughter of Mrs. Dobbs, born just three days earlier:

> It is difficult to judge probabilities of a child as young as that, but it seems to me not quite a normal baby. In its appearance—I should say that perhaps my knowledge of the mother may prejudice me in that regard, but I saw the child at the same time as Mrs. Dobbs' daughter's baby, which is only three days older than this one, and there is a very decided difference in the development of the babies. That was about two weeks ago. There is a look about it that is not quite normal, but just what it is, I can't tell.

This short testimony, and nothing else, formed all the evidence for the crucial third generation of imbeciles. Cross-examination revealed that neither Vivian nor the Dobbs grandchild could walk or talk, and that "Mrs. Dobbs' daughter's baby is a very responsive baby. When you play with it or try to attract its attention—it is a baby that you can play with. The other baby is not. It seems very apathetic and not responsive." Miss Wilhelm then urged Carrie Buck's sterilization: "I think," she said, "it would at least prevent the propagation of her kind." Several years later, Miss Wilhelm denied that she had ever examined Vivian or deemed the child feebleminded.

Unfortunately, Vivian died at age eight of "enteric colitis" (as recorded on her death certificate), an ambiguous diagnosis that could mean many things but may well indicate that she fell victim to one of the preventable childhood diseases of poverty (a grim reminder of the real subject in *Buck* v. *Bell*). She is therefore mute as a witness in our reassessment of her famous case.

When *Buck* v. *Bell* resurfaced in 1980, it immediately struck me that Vivian's case was crucial and that evidence for the mental status of a child who died at age eight might best be found in report cards. I have therefore been trying to track down Vivian Buck's school records for the past four years and have finally succeeded. (They were supplied to me by Dr. Paul A. Lombardo, who also sent other documents, including Miss Wilhelm's testimony, and spent several hours answering my questions by mail and Lord knows how much time playing successful detective in re Vivian's school records. I have never met Dr. Lombardo; he did all this work for kindness, collegiality, and love of the game of knowledge, not for expected reward or even requested acknowledgment. In a profession—academics—so often marred by pettiness and silly squabbling over meaningless priorities, this generosity must be recorded and celebrated as a sign of how things can and should be.)

Vivian Buck was adopted by the Dobbs family, who had raised (but later sent away) her mother, Carrie. As Vivian Alice Elaine Dobbs, she attended the Venable Public Elementary School of Charlottesville for four terms, from September 1930 until May 1932, a month before her death. She was a perfectly normal, quite average student, neither particularly outstanding nor much troubled. In those days before grade inflation, when C meant "good, 81–87" (as defined on her report card) rather than barely scraping by, Vivian Dobbs received A's and B's for deportment and C's for all academic subjects but mathematics (which was always

Vivian Buck (Dobbs) school record for grade 1B, showing satisfactory progress. Note that she was placed on the honor roll in April, 1931. REPRINTED FROM NATURAL HISTORY.

difficult for her, and where she scored D) during her first term in Grade 1A, from September 1930 to January 1931. She improved during her second term in 1B, meriting an A in deportment, C in mathematics, and B in all other academic subjects; she was placed on the honor roll in April 1931. Promoted to 2A, she had trouble during the fall term of 1931, failing mathematics and spelling but receiving A in deportment, B in reading, and C in writing and English. She was "retained in 2A" for the next term—or "left back" as we used to say, and scarcely a sign of imbecility as I remember all my buddies who suffered a similar fate. In any case, she again did well in her final term, with B in deportment, reading, and spelling, and C in writing, English, and mathematics during her last month in school. This daughter of "lewd and immoral" women excelled in deportment and performed adequately, although not brilliantly, in her academic subjects.

In short, we can only agree with the conclusion that Dr. Lombardo has reached in his research on *Buck* v. *Bell*—there were no imbeciles, not a one, among the three generations of Bucks. I don't know that such correction of cruel but forgotten errors of history counts for much, but I find it both symbolic and satisfying to learn that forced eugenic sterilization, a procedure of such dubious morality, earned its official justification (and won its most quoted line of rhetoric) on a patent falsehood.

Carrie Buck died last year. By a quirk of fate, and not by memory or design, she was buried just a few steps from her only daughter's grave. In the umpteenth and ultimate verse of a favorite old ballad, a rose and a brier—the sweet and the bitter—emerge from the tombs of Barbara Allen and her lover, twining about each other in the union of death. May Carrie and Vivian, victims in different ways and in the flower of youth, rest together in peace.

21 | Singapore's Patrimony (and Matrimony)

SOME HISTORICAL ARGUMENTS are so intrinsically illogical or implausible that, following their fall from grace, we do not anticipate any subsequent resurrection in later times and contexts. The disappearance of some ideas should be as irrevocable as the extinction of species.

Of all invalid notions in the long history of eugenics—the attempt to "improve" human qualities by selective breeding—no argument strikes me as more silly or self-serving than the attempt to infer people's intrinsic, genetically based "intelligence" from the number of years they attended school. Dumb folks, or so the argument went, just can't hack it in the classroom; they abandon formal education as soon as they can. The fallacy, of course, lies in a mix-up, indeed a reversal, of cause and effect. We do not deny that adults who strike us as intelligent usually (but by no means always) spent many years in school. But common sense dictates that their achievements are largely a result of the teaching and the learning itself (and of the favorable economic and intellectual environments that permit the luxury of advanced education), not of a genetic patrimony that kept them on school benches. Unless education is a monumental waste of time, teachers must be transmitting, and students receiving, something of value.

This reversed explanation makes such evident sense that even the staunchest of eugenicists abandoned the original genetic version long ago. The genetic argument was quite

319

popular from the origin of IQ testing early in our century until the mid-1920s, but I can find scarcely any reference to it thereafter—although Cyril Burt, that great old faker and discredited doyen of hereditarians, did write in 1947:

> It is impossible for a pint jug to hold more than a pint of milk; and it is equally impossible for a child's educational attainments to rise higher than his educable capacity permits.

In my favorite example of the original, genetic version, Harvard psychologist R.M. Yerkes tested nearly two million recruits to this man's army during World War I and calculated a correlation coefficient of 0.75 between measured intelligence and years of schooling. He concluded:

> The theory that native intelligence is one of the most important conditioning factors in continuance in school is certainly borne out by this accumulation of data.

Yerkes then noted a further correlation between low scores of blacks on his tests and limited or absent schooling. He seemed on the verge of a significant social observation when he wrote:

> Negro recruits though brought up in this country where elementary education is supposedly not only free but compulsory on all, report no schooling in astonishingly large proportion.

But he gave the data his customary genetic twist by arguing that a disinclination to attend school can only reflect low innate intelligence. Not a word did he say about the poor quality (and budgets) of segregated schools or the need for early and gainful employment among the impoverished. (Ashley Montagu reexamined Yerkes's voluminous data twenty years later and, in a famous paper, showed that blacks in several northern states with generous school budgets and strong commitments to education tested better than whites in southern states with the same years of schooling. I could almost hear the old-line eugenicists sputtering from their graves, "Yes, but, but only the most intelligent blacks were smart enough to move north.")

I did not, in any case, ever expect to see Yerkes's argument revived as a hereditarian weapon in the ongoing debate about human intelligence. I was wrong. The reincarnation is particularly intriguing because it comes from a place and culture so distant from the original context of IQ testing in Western Europe and America. It should teach us that debates among academics are not always the impotent displays of arcane mental gymnastics so often portrayed in our satires and stereotypes, but that ideas can have important social consequences with impacts upon the lives of millions. Old notions may emerge later, often in curiously altered contexts, but their source can still be recognized and traced to claims made in the name of science yet never really supported by more than the social prejudices (often unrecognized) of their proposers. Ideas matter in tangible ways.

I recently received from some friends in Singapore a thick package of xeroxed reports from the English-language press of their nation. These pages covered a debate that has raged in their country since August 1983, when in his annual National Day Rally speech (an equivalent to our "state of the union" message, I gather), Prime Minister Lee Kwan Yew abandoned his customary account of economic prospects and progress and, instead, devoted his remarks to what he regards as a great danger threatening his nation. The headline of the *Straits Times* for August 15 read (Singapore was once the primary city of a British colony named Straits Settlement): "Get Hitched . . . and don't stop at one. PM sees depletion of talent pool in 25 years unless better educated wed and have more children."

Prime Minister Lee had studied the 1980 census figures and found a troubling relationship between the years that women spend in school and the number of children subsequently born. Specifically, Mr. Lee noted that women with no education have, on average, 3.5 children; with primary education, 2.7; with secondary schooling, 2.0; and with university degrees, only 1.65. He stated:

> The better educated the people are, the less children they have. They can see the advantages of a small family. They know the burden of bringing up a large family. . . . The better educated the woman is, the less children she has.

So far, of course, Prime Minister Lee had merely noted for his nation a demographic pattern common to nearly every modern technological society. Women with advanced degrees and interesting careers do not wish to spend their lives at home, bearing and raising large families. Mr. Lee acknowledged:

> It is too late for us to reverse our policies and have our women go back to their primary role as mothers. . . . Our women will not stand for it. And anyway, they have already become too important a factor in the economy.

But why is this pattern troubling? It has existed for generations in many nations, our own for example, with no apparent detriment to our mental or moral stock. The correlation of education with fewer children becomes a dilemma only when you infuse Yerkes's old and discredited argument that people with fewer years of schooling are irrevocably and biologically less intelligent, and that their stupidity will be inherited by their offspring. Mr. Lee proposed just this argument, thus setting off what Singapore's press then dubbed "the great marriage debate."

The prime minister is not, of course, unaware that years in school can reflect economic advantages and family traditions with little bearing on inherited smarts. But he made a specific argument that deemphasized to insignificance the potential contribution of such environmental factors to years of schooling. Singapore has made great and recent advances in education: universal schooling was introduced during the 1960s and university places were opened to all qualified candidates. Before these reforms, Lee argued, many genetically bright children grew up in poor homes and never received an adequate education. But, he contends, this single generation of universal opportunity resolved all previous genetic inequities in one swoop. Able children of poor parents were discovered and educated to their level of competence. Society has sorted itself out along lines of genetic capacity—and level of education is now a sure guide to inherited ability.

We gave universal education to the first generation in the early 1960s. In the 1960s and '70s, we reaped a big crop of able boys and girls. They came from bright parents, many of whom were never educated. In their parents' generation, the able and not-so-able both had large families. This is a once-ever bumper crop which is not likely to be repeated. For once this generation of children from uneducated parents have received their education in the late 1960s and '70s, and the bright ones make it to the top, to tertiary [that is, university] levels, they will have less than two children per ever-married woman. They will not have large families like their parents.

Lee then sketched a dire picture of gradual genetic deterioration:

If we continue to reproduce ourselves in this lopsided way, we will be unable to maintain our present standards. Levels of competence will decline. Our economy will falter, the administration will suffer, and the society will decline. For how can we avoid lowering performance, when for every two graduates (with some exaggeration to make the point), in 25 years' time there will be one graduate, and for every two uneducated workers, there will be three?

So far, I have not proved my case—that the worst arguments raised by hereditarians in the great nature-nurture wars of Western intellectuals can resurface with great social impact in later and quite different contexts. Mr. Lee's arguments certainly sound like a replay of the immigration debate in America during the early 1920s or of the long controversy in Britain over establishing separate, state-supported schools (done for many years) for bright and benighted children. After all, the arguments are easy to construct, however flawed. Perhaps the prime minister of Singapore merely devised them anew, with no input from older, Western incarnations.

But another key passage in Lee's speech—the one that set off waves of recognition and inspired me to write this essay —locates the source of Lee's claims in old fallacies of the Western literature. I have left one crucial part of the argument out—the "positive" justification for a predominance of heredity in intellectual achievement (versus the merely negative claim that universal education should smooth out any environmental component). Lee stated, in a passage that sent a frisson of *déjà-vu* up my spine:

> A person's performance depends on nature and nurture. There is increasing evidence that nature, or what is inherited, is the greater determinant of a person's performance than nurture (or education and environment). . . . The conclusion the researchers draw is that 80 percent is nature, or inherited, and 20 percent the differences from different environment and upbringing.

Note the giveaway phrase: "80 percent" (supplemented by Lee's specific references to studies of identical twins reared apart). All cognoscenti of the Western debate will immediately recognize the source of this claim in the "standard figure" so often cited by hereditarians (especially by Arthur Jensen in his notorious 1969 article entitled "How Much Can We Boost IQ and Scholastic Achievement") that IQ has a measured heritability of 80 percent.

The fallacies of this 80 percent formula, both of fact and interpretation, have also been thoroughly aired back home, but this aspect of the debate has, alas, apparently not penetrated to Singapore.

When Jensen advocated an 80 percent heritability, his primary defense rested upon Cyril Burt's study of identical twins separated early in life and raised apart. Burt, the grand old man of hereditarianism, wrote his first paper in 1909 (just four years after Binet published his initial IQ test) and continued, with steadfast consistency, to advance the same arguments until his death in 1971. His study of separated twins won special fame because he had amassed so large a sample for this rarest of all animals—more than fifty cases—where no previous researcher had managed to find even half so many. We now know that Burt's "study" was perhaps the most spectacular case of outright scientific fraud in our century—no problem locating fifty pairs of separated twins when they exist only in your own head.

Burt's hereditarian supporters first reacted to the charge of fraud by attributing the accusation to left-wing environmentalist ideologues out to destroy a man by innuendo when they couldn't overwhelm him by logic or evidence. Now that Burt's fraud has been established beyond any possible doubt (see L.S. Hearnshaw's biography, *Cyril Burt, Psychologist*), his erstwhile supporters advance another argument—the 80 percent figure is so well established from other studies that Burt's "corroboration" didn't matter.

In my reading, the literature on estimates of heritability for IQ is a confusing mess—with values from 80 percent, still cited by Jensen and others, all the way down to Leon Kamin's contention (see bibliography) that existing information is not incompatible with a true heritability of flat zero. In any case, the actual number hardly matters, for Lee's argument rests upon a deeper and more basic fallacy —a false interpretation of what heritability means, whatever its numerical value.

The problem begins with a common and incorrect equation of heritable with "fixed and inevitable." Most people, when they hear that IQ has a heritability of 80 percent, conclude that four-fifths of its value is irrevocably set in our genes with only one-fifth subject to improvement by good education and environment. Prime Minister Lee fell right into this old trap of false reason when he concluded that 80 percent heritability established the predominance of nature over nurture.

Heritability, as a technical term, measures how much variation in the appearance of a trait within a population (height, eye color, or IQ, for example) can be accounted for by genetic differences among individuals. Heritability simply isn't a measure of flexibility or inflexibility in the potential expression of a trait. A type of visual impairment, for example, might be 100 percent heritable but still easily corrected to normal vision by a pair of eyeglasses. Even if IQ were 80 percent heritable, it might still be subject to major improvement by proper education. (I do not claim that all heritable traits are easily altered; some inherited visual handicaps cannot be overcome by any available technology. I merely point out that heritability is not a measure of intrinsic and unchangeable biology.) Thus, I confess I have never been much interested in the debate over IQ's heritability—for even a very high value (which is far from established) would not speak to the main issue, so accurately characterized by Jensen in the title of his article—how much can we boost IQ and scholastic achievement? And I haven't even mentioned (and won't discuss, lest this essay become interminable) the deeper fallacy of this whole debate—the assumption that so wonderfully multifarious a notion as intelligence can be meaningfully measured by a single number, with people ranked thereby along a unilinear scale of mental worth. IQ may have a high heritability, but if this venerable measure of intelligence is (as I suspect) a meaningless abstraction, then who cares? The first joint of my right ring finger probably has a higher heritability than IQ, but no one bothers to measure its length because the trait has neither independent reality nor importance.

In arguing that Prime Minister Lee has based his fears for Singapore's intellectual deterioration upon a false reading of some dubious Western data, I emphatically disclaim any right to pontificate about Singapore's problems or their potential solutions. I am qualified to comment on Mr. Lee's nation only by the first criterion of the old joke that experts on other countries have lived there for either less than a week or more than thirty years. Nonetheless, buttinsky that I am, I cannot resist two small intrusions. I question, first, whether a nation with such diverse cultural traditions among its Chinese, Malay, and Indian sectors can really expect to even out all environmental influences in just one generation of educational opportunity. Second, I wonder whether the world's most densely populated nation (excluding such tiny city-states as Monaco) should really be encouraging a higher reproductive rate in any segment of its population. Despite my allegiance to cultural relativism, I still maintain a right to comment when other traditions directly borrow my own culture's illogic.

The greatest barrier to understanding the real issue in this historical debate may best be expressed by exposing the false approach encouraged by that euphonious contrast of supposed opposites—*nature* and *nurture*. (How I wish that English did not contain such an irresistible pair—for language channels thought, often in unfortunate directions. In previous centuries, the felicity of phrase underscoring a comparison between God's *words* and his *works* encouraged a misreading of nature as a mirror of biblical truth. In our times, an imagined antithesis of nature and nurture provokes a compartmentalization quite foreign to our world of interactions.) All complex human traits are built by an inextricable mixture of varied environments working upon the unfolding of a program bound in inherited DNA. Interaction begins at the moment of fertilization and continues to the instant of death; we cannot neatly divide any human behavior into a part rigidly determined by biology and a portion subject to change by external influence.

The real issue is *biological* potentiality versus *biological* determinism. We are all interactionists; we all acknowledge the powerful influence of biology upon human behavior. But determinists, like Arthur Jensen and Prime Minister Lee (at least in his August speech), use biology to construct a *theory of limits*. In Mr. Lee's version, lack of schooling implies ineradicable want of intelligence since the fault (or at least four-fifths of it) does indeed lie not in our stars but in ourselves if we are underlings. Potentialists acknowledge the importance of biology but stress that complexities of interaction, and the resultant flexibility of behavior, preclude rigid genetic programming as the basis for human achievement.

Biological determinism has a long-standing (and continuing) political use as a tool for justifying the inequities of a status quo by blaming the victim—as John Conyers, Jr., one of our few black congressmen, states in a powerful Op-Ed piece in the *New York Times* on December 28, 1983. Conyers begins:

> In the 1950s, much of the sociological literature on poverty attributed the economic plight of blacks and other minorities to what it said was inherent laziness and intellectual inferiority. This deflected attention from the virtually insurmountable walls of segregation that blocked social and economic mobility.

Conyers then analyzes a growing literature that seeks genetic causes for high mortality rates among blacks, particularly for various forms of cancer. "In the workplace," Conyers writes,

> blacks have a 37 percent higher risk of occupationally induced disease and a 20 percent higher death rate from occupationally related diseases.

Susceptibility to disease may be influenced by genetic constitution, and racial groups may vary in their average propensities. But if we focus on unsupported speculations about inheritance, we neglect the immediate root in racism and economic disadvantage—for these pervasive problems are surely major causes of the discrepancy, which could then be reduced or eliminated by social reform. (As an obvious political comment, location of the cause in intractable biology decreases pressure for the same reforms.) Conyers continues:

> Just as in the 1950s, blacks are being told that their problems are largely self-inflicted, that their poor

health is a manifestation of immoderate personal habits. Such blame-the-victim strategies . . . serve to divert attention from the fact that blacks are the targets of a disproportionate threat from toxins both in the workplace, where they are assigned the dirtiest and most hazardous jobs, and in their homes, which tend to be situated in the most polluted communities.

As an example, Conyers notes that black steelworkers in coke plants display twice the cancer death rate of white workers, with eight times the white rate for lung cancer, in particular. "This disparity," Conyers argues,

is explainable by job patterns: 89 percent of black workers labor at coke ovens—the most dangerous part of the industry; only 32 percent of their white co-workers do.

Shall we strive directly to improve working conditions or speculate about inherent racial differences? Even if we prefer genetic hypotheses, we could only test them by equalizing (and improving) our workplaces, and then assessing the impact upon mortality. Similarly, should we proclaim that women with little schooling must be intractably stupid or should we remove social and economic obstacles, push universal education a little bit harder, and see how well these women do? In the midst of Singapore's great marriage debate, the *Jakarta Post* peeked in on its neighbor's brouhaha and commented: "It would be more sensible and less controversial to build more schools."

Postscript

The situation in Singapore has become, in Alice's immortal words, "curiouser and curiouser" since I wrote this essay. Some reports seem almost comical, but we laugh at our

peril (as I shall soon document). Soon after Prime Minister
Lee Kwan Yew's speech and the resulting furor described
in my original essay, Deputy Prime Minister Dr. Goh Keng
Swee unveiled his first package of countermeasures to the
public. They included the establishment of computer dating
services to foster eugenically appropriate matches and instructions to the National University of Singapore that undergraduate courses on courtship be introduced in order to
hone the skills of shy but able potential breeders. According
to the *New York Times* (February 12, 1984), Singapore's
state-owned television "is planning to run a drama series
that will seek to show that unmarried career women are
incomplete and that their lives are void."

More seriously, and verging on the insidious, Prime Minister Lee has now instituted the first official measures of
preference and incentive. The Family Planning Board of
Singapore has reversed its long-standing campaign of persuasion for restriction of families to two children—but only
in propaganda directed towards the well educated. The
Board is now pursuing a "dual-message" campaign: "Graduates and professionals will be told to go forth and multiply;
the less educated will be urged to have no more than two
children" (*New York Times,* February 12, 1984).

As a first explicit act, the Government proclaimed in January, 1984, that women with university degrees will be
awarded priority for enrolling children in the primary
schools of their choice. The less educated—now get this
and shudder—will get next preference if they agree to sterilization after birth of their first or second child (*New York
Times,* February 12, 1984).

Prime Minister Lee's plans have not met with universal
approbation, either in Singapore or in neighboring countries. Warren Y. Brockelman of Mahidol University in Bangkok joined Yongyuth Yuthavong and ten other members of
the Faculty of Science in a vigorous protest (published in
the *Bangkok Post* for February 16, 1984. I thank Dr. Brockelman for sending, via David Woodruff, the documents that
I used to write this postscript). They write:

There is no evidence that birth rate differentials between economic classes or educational levels produce any changes in the genetic structure of a human population . . . . A particularly counterproductive and unfair aspect of the new policy is that children born to well educated mothers will be given preference over others in school admissions. The effect of this policy will be to ensure that less educated families remain uneducated and retain high birth rates. It will not increase the pool of educated talent. A more sensible policy would be to give the children of less educated couples preference in admissions, so that they will rise in socio-economic achievement level and attain the lower birth rates usually associated with such achievement.

In neighboring Malaysia, Chee Heng Leng and Chan Chee Khoon have published a series of critiques inspired by Singapore's renewal of eugenics (*Designer Genes, IQ, Ideology and Biology,* INSAN, Selangor, Malaysia—the cover features a photo of a pair of denims—Lee brand, natch. I am pleased that they were able to include an essay of mine, from *Ever Since Darwin,* in their collection). Drs. Chee and Chan point out that similar ideas are afoot in Malaysia (though not yet translated into official policy), where Prime Minister Datuk Seri Dr. Mahathir Mohamad has argued that native Malays have inherited a weak and easygoing character as a result of their genial physical environment combined with inbreeding (while ethnic Chinese are a hardier lot bred in a tougher land). Chee and Chan sum up the situation in Singapore admirably:

What is remarkable about the current Singaporean situation is really the crude way in which the "heritability of IQ" concept has been formulated. Furthermore, so-called scientific data that lost all credibility in scientific circles a decade ago are being used to buttress these assertions . . . The Singapore situation is

also amazing in that these "scientific" pronouncements have been rapidly translated into social policies,
which blatantly favor the upper class and discriminate
against the poor majority of the Singaporean population.

6 | Darwiniana

22 | Hannah West's Left Shoulder and the Origin of Natural Selection

IN HIS ESSAY "Technical Education," written in 1877, Thomas Henry Huxley proclaimed that "the great end of life is not knowledge but action." Since Huxley was no intellectual slouch, we may be confident that he was not advocating thoughtless exertion, but arguing that hard-won knowledge only gains its highest value in utility. As Marx wrote in his last thesis on Feuerbach: "Philosophers have thus far only interpreted the world in various ways; the point, however, is to change it."

Pristine originality is an illusion; all great ideas were thought and expressed before a conventional founder first proclaimed them. Copernicus did not reverse heavenly motion single-handedly, and Darwin did not invent evolution. Conventional founders win their just reputations because they prepare for action and grasp the full implication of ideas that predecessors expressed with little appreciation of their revolutionary power.

All scholars know that several prominent scientists—Lamarck in particular—developed elaborate systems of evolutionary thought before Darwin. Many, however, suppose that Darwin was the true originator of his own particular theory about *how* evolution occurred—natural selection. Yet, by his own belated admission (in the historical preface added to later editions of the *Origin of Species*), Darwin allowed that two authors had preceded him in formulating the principle of natural selection. He also argued, at least by implication—and I heartily agree—that neither of these anticipations diluted his claim to fame or originality. He had not initially disregarded them through ill will, but simply because he had never heard about them, despite his thoroughly omnivorous habits of reading and correspondence. The reasons for this justifiable ignorance reinforce Darwin's status and help us to understand the difference between merely stating an idea and understanding what it can do and mean.

One of Darwin's predecessors, the Scottish naturalist and fruit grower Patrick Matthew, published his version of natural selection in 1831 as an appendix to a work entitled *Naval Timber and Arboriculture*. And there it languished, unnoticed in its odd context, until Darwin published in 1859. Matthew then wrote a letter to the *Gardeners' Chronicle* asserting his priority not only for natural selection but also for

the first proposal of the steam ram (also claimed since by several others—English, French, and American) and a navy of steam gun-boats as requisite in future maritime war, and which, like the organic selection law, are only as yet making way.

Darwin responded to the *Gardeners' Chronicle* on April 21, 1860 (I thank W.J. Dempster for sending me copies of this correspondence and for urging my attention to Matthew's views):

I have been much interested by Mr. Patrick Matthew's communication in the Number of your Paper, dated April 7. I freely acknowledge that Mr. Matthew has anticipated by many years the explanation which I have offered of the origin of species, under the name of natural selection. I think that no one will feel surprised that neither I, nor apparently any other naturalist, has heard of Mr. Matthew's views, considering how briefly they are given, and that they appeared in the Appendix to a work on Naval Timber and Arboriculture. I can do no more than offer my apologies to Mr.

Matthew for my entire ignorance of his publication. If another edition of my work is called for, I will insert a notice to the foregoing effect.

The second, and earlier, anticipation of natural selection was not presented in so obscure a context. In 1813, William Charles Wells, another Scottish scientist and physician (though born in Charleston, South Carolina), delivered a paper before the Royal Society of London, England's pre-eminent institution of science. It bore one of those wonder-fully extended titles so common at the time: *Account of a female of the white race of mankind, parts of whose skin resembles that of a negro, with some observations on the causes of the differences in color and form between the white and negro races of men.*

The essay made no recorded impact at its presentation, and Wells did not print it at the time. As he lay dying of heart disease five years later, Wells prepared for publication a single volume of his more important essays. This volume, published posthumously in 1818, included the short 1813 address almost as an afterthought at the very end. Wells's volume was well enough received, for it featured the two essays that had won his secure, if limited, fame—one on the formation of dew (a problem solved definitively by Wells, who proved that dew is neither invisible rain nor an exuda-tion of plants, but a result of condensation from surround-ing air), and another on why our two eyes see but a single image. Ironically, and as testimony to the total obscurity of Wells's short essay on the origin of human skin color, when Hugh Falconer proposed Darwin for the Copley Medal of the Royal Society in 1864, he praised Darwin by comparing his methods of research with those followed in Wells's ex-cellent treatise on dew: "It may be compared with Dr. Wells's 'Essay on Dew' as original, exhaustive and complete —containing the closest observation with large and impor-tant generalization." Falconer apparently never realized that the volume he had consulted to read Wells's "Essay on Dew" also contained an anticipatory statement about natu-ral selection itself.

Wells was an austere, intensely private, idiosyncratic

man. He had, by his own account, few friends, fewer patients, and very little cash (largely because he spent most of his life repaying loans to his few good friends). He lived his adult life alone in London. He never married, socialized little, and published less. The autobiography prefixed to his volume of essays bemoans his persistent financial straits, particularly his inability to maintain a carriage, thereby foreclosing most social activity and access to potential patients (in those happy but bygone days when home visits by physicians were practically mandatory).

Although born in America, Wells was the son of intense British loyalists. Wells recorded his father's anxiety that a young man might be won to the republican cause in an agitated prerevolutionary America:

> He, fearing that I should become tainted with the disloyal principles which began immediately after the peace of 1763 to prevail throughout America, obliged me to wear a tartan coat, and a blue Scotch bonnet, hoping, by these means, to make me consider myself a Scotchman. The persecution I hence suffered produced this effect completely.

Wells had little good to say for America, for he attributed to his early life in South Carolina virtually all the faults of his later days, including this embarrassed admission:

> What I shall next say will no doubt be held very ridiculous. I lived till I was near 11 years old, close upon the harbor of a large sea-port in America, and by this means associated much with blackguard sailor boys. To this I attribute a practice of swearing, of which I have from the time of being a child, been frequently guilty, when my feelings have been agitated, and even sometimes when no excuse of this kind has existed.

Wells was therefore happy to leave America for an education in Britain. He returned to Carolina (then in Royalist hands) in 1781 to look after his father's affairs; but he ended up in jail after the overturn of political power and was only too happy to win repatriation, this time permanently, to Britain. He moved to London and was licensed by the Royal College of Physicians in 1788. His 1792 essay on single vision with two eyes secured his election to the Royal Society, while his 1814 essay on dew won him the society's coveted Rumford Medal. Despite the quality and renown of these works, Wells published little else. His autobiography does not even mention the 1813 essay on human skin color, and we have no indication that he afforded it any significance in his own mind or that he recognized any further-ranging implications for its ideas.

Like so many general statements written by physicians, Wells's 1813 essay on natural selection begins with a description of an unusual medical case history. Hannah West, a young woman from Sussex and the daughter of "a footman in a gentleman's family," visited him for observation of her peculiar skin. Her parents and all relatives were conventional Caucasians, but Hannah West, although appropriately pale skinned everywhere else, was "as dark as any negro" on her left shoulder, arm, forearm, and hand. In deference to the venerable theory, then still prevalent, of "maternal impressions" (see last essay in *Hen's Teeth and Horse's Toes*), West's family and neighbors attributed her affliction to the following peculiar event:

> Her mother . . . received a fright, while pregnant with her, by accidently treading on a live lobster; and to this was attributed the blackness of part of her skin, which was observed at her birth.

Wells observed Hannah West carefully, noted the sharp transition between her unusual dark and expected white skin, and marveled at the blackness of her left arm—"darker than the corresponding part in any negro whom I have seen; for the palm of her hand and inside of her fingers are black, whereas these parts in a negro are only of a tawny hue." But, in truth, Wells never transcended the purely descriptive and reported nothing of general interest. Even the basic premise of his account was erroneous; whites with large patches of melanic skin bear no meaningful resemblance, genealogical or otherwise, to black people. Had Wells not appended seven pages of speculation on the origin of human skin colors to his report, it would surely have fallen into total and permanent oblivion, rather than mere obscurity (with later resurrection as a curiosity). These seven pages, the afterthought to an essay published as an afterthought, include a two- to three-page section on natural selection, the first clear and recognized statement of Darwin's great principle.

Wells begins diffidently, fearful perhaps that too much speculation will dilute the value of his sober observations on Hannah West's unusual skin:

> On considering the difference of color between Europeans and Africans, a view has occurred to me of this subject, which has not been given by any author, whose works have fallen into my hands. I shall, therefore, venture to mention it here, though at the hazard of its being thought rather fanciful than just.

Wells invokes natural selection to explain the success of black people in hot climates. Beginning with the usual and unstated racist assumption that white skin is proper and primary, Wells imagines that the original inhabitants of Africa were lighter than their current descendants. He explains the change by natural selection and even invokes Darwin's favorite argument of analogy with artificial selection as practiced by animal breeders:

> Those who attend to the improvement of domestic animals, when they find individuals possessing, in a greater degree than common, the qualities they desire, couple a male and female of these together, then take the best of their offspring as a new stock, and in this way proceed, till they approach as near the point in view, as the nature of things will permit. But, what

is here done by art, seems to be done, with equal efficacy, though more slowly, by nature, in the formation of varieties of mankind, fitted for the country which they inhabit.

This statement has been cited in several previous works (it is the "standard" quote from Wells), but I do not think that any previous commentator has recognized the decidedly unorthodox character of Wells's presentation. Curiously, its unusual features presage some of the arguments now being presented by many evolutionists against the strict construction of Darwinism that has been so popular during the past twenty years or so.

The conventional, strictly Darwinian, argument would feature direct adaptation of skin color and evolutionary change driven by competition among individual organisms. In other words, we would argue that black skin offered direct advantages in hot climates and that it arose by differential survival and propagation of darker individuals within a population. Wells explicitly denies both parts of this scenario.

For adaptation, Wells rebuts the idea that black skin provides any benefit in itself (he was probably wrong) and claims instead that some other physiological feature adapts black people to hot climates by conferring resistance to tropical diseases. Wells speculates that this feature may be subtle and not manifested in evident morphology. Black skin may be correlated with this feature for some unknown developmental reason and may therefore serve as a signpost of advantage, while providing no direct benefit itself:

I do not, however, suppose, that their different susceptibility of diseases depends, properly, on their difference of color. On the contrary, I think it probable, that this is only a sign of some difference in them, which, though strongly manifested by its effects in life, is yet too subtle to be discovered by an anatomist after death; in like manner as a human body, which is incapable of receiving the small-pox, differs in no observable thing from another, which is still liable to be affected with that disease.

Darwin puzzled over these "correlations of growth" and recognized that many features may offer no direct benefit, yet still characterize large groups by their forced physiological relationship with other traits. Extreme versions of Darwinism have forgotten this subtlety and have tried to find direct adaptive advantages, often by purely speculative argument, for nearly every widespread feature.

Wells's theme of "nonadaptive consequences" has been reasserted of late in an atmosphere of renewed attention to developmental patterns and the integrity of organization (animals cannot be analyzed as an amalgam of independent parts). Wells's treatment of color follows these recent criticisms.

For selection, the usual argument would propose a human population with considerable variation in skin color among its members. People with dark skin would, on average, be more successful in raising offspring and, slowly but surely, skin color within the population would shift toward darker hues. In other words, evolutionary change occurs by competition among individuals *within* a population (the "struggle for existence").

Wells explicitly denies this usual form of selection by claiming that favorable variants cannot spread through large and stable populations. His argument is incorrect and based on a false view of heredity, then current, called blending inheritance—the idea that all favorable variants will be diluted by half in offspring through intermarriage with a normal member of the population. The diluted offspring will then usually marry a normal individual (since favorable variants are so rare) and the subsequent generation will be diluted to one-quarter. Soon, the rare and favorable variant will be entirely swamped out. Heredity doesn't work this way (although Wells couldn't know what Mendel would discover fifty years later). Favorable traits often arise by mutation, and such features cannot be diluted by breeding with normal individuals. The mutation (if recessive) may not be expressed in the next generation, but it will not be eliminated. Wells's belief in blending inheritance led him to deny selection by slow transformation within a population:

> Those varieties [that is, favorable variations], for the most part, quickly disappear, from the intermarriages of different families. Thus, if a very tall man be produced, he very commonly marries a woman much less than himself, and their progeny scarcely differs in size from their countrymen.

How then can selection work? Wells argues that favorable variants can spread, presumably by chance rather than by selection (although he is not explicit), through small and mobile populations where vast numbers of normal individuals cannot impose dilution by backbreeding:

> In districts, however, of very small extent, and having little intercourse with other countries, an accidental difference in the appearance of the inhabitants will often descend to their late posterity.

Thus, Wells conjectures that the people of Africa were initially divided into tiny, noninteracting populations. By chance, different average colors (and accompanying resistances to disease) became established among these populations. Selection then acted by competition between populations already different (for reasons unrelated to natural selection) in average skin color. Within each group, color was relatively constant and selection could only operate by sorting the groups themselves. Selection, in other words, occurs between groups, not among individuals within a group.

> Of the accidental varieties of man [meaning populations this time; Wells and his contemporaries used the word *variety* both for distinct individuals and for different populations], which would occur among the first

few and scattered inhabitants of the middle regions of Africa, some one would be better fitted than the others to bear the diseases of the country. This race would consequently multiply, while the others would decrease, not only from their inability to sustain the attacks of disease, but from their incapacity of contending with their more vigorous neighbors. The color of this vigorous race I take for granted, from what has been already said, would be dark.

The locus or level of selection is a "hot topic" within evolutionary theory today. While no one denies that selection works powerfully at the traditional level of differences among organisms within a population, other modes may also be important. The idea that selection may operate primarily *among* local populations—so-called intergroup selection—has long been advocated by the great geneticist Sewall Wright (who still argues eloquently for his position at age 95). Wright's eclipse within strict Darwinian circles has recently been reversed and intergroup selection is now receiving a more favorable second look. I find it intriguing that the first formulation of natural selection advocated an intergroup process rather than the traditional focus on competing organisms.

Nonetheless, although Wells's argument was unorthodox by later Darwinian standards, it surely states the principle of natural selection. We must therefore return to our initial question. Why were these Darwinian harbingers totally ignored, and why does Darwin deserve his status (and, I fear, Wells and Matthew theirs as well).

Loren Eiseley makes the astute point (in *Darwin's Century*) that Wells's argument, as stated, cannot be extended or generalized to yield the panoply of evolutionary change through life's history. Everyone knew that organisms varied and that local breeds could be manufactured from this raw material (how else did dog and pigeon fanciers work, not to mention farmers). But variation among breeds of dogs does not automatically extrapolate to the transformation from fish to human. Perhaps species have fixed and God-given limits of variation. We may make new breeds by selecting for extremes within these limits, but we cannot transcend the boundary to construct fundamentally new creatures. Wells does not generalize his argument to encompass large-scale evolutionary change, and hindsight alone may permit us to read his speculations as harbingers to Darwin's overturn of biology.

Still, Wells's failure to generalize cannot be the main reason for his obscurity. (Darwin, by the way, only learned of Wells's work from an American correspondent with antiquarian bibliographic interests.) The primary rationale is uncomplex. Ideas are cheap; mere statement counts for little or nothing. Intellectual fame accrues to people with the vision to make a good idea work in two ways: by using it to make new discoveries and by recognizing its implications as a far-ranging instrument for transforming general attitudes.

We have no reason to suspect that either Wells or Matthew recognized any of the revolutionary power behind his cleverness. Wells presented natural selection as an appendage to an essay he didn't even bother to publish until he lay dying. Matthew buried it among his trees and saw no forest (although he, unlike Wells, did advocate evolution as the cause of life's history). Indeed, in a second letter responding to Darwin's apology in 1860, Matthew damns Darwin with faint praise (and inadvertently condemns himself) by arguing that he never made much of selection because, unlike Darwin who struggled so hard to formulate the principle, he had grasped it as an evident deduction from the nature of things. He regarded it as necessarily true, almost trivial in that sense, and thus unworthy of much development. Matthew therefore missed all its significance:

> To me the conception of this law of Nature came intuitively as a self-evident fact, almost without an effort of concentrated thought. Mr. Darwin here seems to have more merit in the discovery than I have had—to me it did not appear a discovery. He seems to have worked it out by inductive reason, slowly and with due caution

to have made his way synthetically from fact to fact onwards; while with me it was by a general glance at the scheme of Nature that I estimated this select production of species as an a priori recognizable fact—an axiom, requiring only to be pointed out to be admitted by unprejudiced minds of sufficient grasp.

Darwin, on the other hand, used natural selection as the intellectual fulcrum of an entire career. He interpreted human evolution in its light, reformulated the principles of psychology, and explained the coevolution of orchids and their pollinating insects, the biogeographic distribution of organisms, the habits and actions of worms—a rich panoply of issues from the largest enigmas of life to the smallest quirks of particular organisms. He established a workable research program for an entire profession.

I have found no documents of human thought more exciting than the notebooks that Darwin filled in London as a young man in his late twenties, just returned from five years aboard the *Beagle*. He had the key to a new view of life, and he knew it. His mind ranged over the entire intellectual landscape, from biology to psychology, morality, philosophy, and literature. Evolution by natural selection impinged on everything. Wells and Matthew had stated the same principle, but they had then either forgotten or had failed to draw any implications. Darwin sat in London, a young man rebuilding a world of thought. Consider but one statement, as a symbol of his achievement and a fitting end to this essay. Charles Darwin, cutting through two thousand years of tradition in Western philosophy with one epigrammatic note to himself:

Plato says in *Phaedo* that our "imaginary ideas" arise from the preexistence of the soul, are not derivable from experience—read monkeys for preexistence.

23 | Darwin at Sea—and the Virtues of Port

CHARLES DARWIN AND ABRAHAM LINCOLN were born on the same day—February 12, 1809. They are also linked in another curious way—for both must simultaneously play, and for similar reasons, the role of man and legend. In a nation too young for mythic heroes, flesh and blood must substitute. Hence schoolchildren learn about Honest Abe, who freed the slaves single-handedly for simple justice, and who, as a young man, trudged for miles to return a few cents to a woman he had inadvertently short-changed. This legendary Lincoln may fulfill a national or psychological need, but historians must also labor to rescue the real, and wondrously complex, man from such a factually inaccurate role. Similarly, science worships no gods, and ancient sages are in strictly short supply. Historical figures must again form the stuff of needed legends. The apple beans Newton; Galileo drops his missiles from the Leaning Tower; and Darwin, alone at sea, transforms the intellectual world in splendid mental isolation.

The myth of the *Beagle*—that Darwin became an evolutionist by simple, unbiased observation of an entire world laid out before him during a five-year voyage around the world—fits all our romantic criteria for the best of legends: a young man, freed from the trammels of English society and its constraining presuppositions, face to face with nature, parrying his fresh and formidable mind with all the challenges provided by plants, animals, and rocks through-

347

out the globe. He leaves England in 1831, planning to become a country parson upon his return. He returns in 1836, having seen evolution in the raw, understanding (albeit dimly) its implications and committed to a scientific life as evolutionary thinker. The chief catalyst: the Galápagos Islands. The main actors: tortoises, mockingbirds, and above all, thirteen species of Darwin's finches—the finest evolutionary laboratory offered to us by nature.

We may need simple and heroic legends for that peculiar genre of literature known as the textbook. But historians must also labor to rescue human beings from their legends in science—if only so that we may understand the process of scientific thought aright. Darwin, to begin, did not become an evolutionist until several months after his return to London—probably not until March 1837 (the *Beagle* docked in October 1836). He did not appreciate the evolutionary significance of the Galápagos while he was there, and he originally misunderstood the finches so thoroughly that he was barely able to reconstruct the story later from his sadly inadequate records. The legend of the finches may persist, but it has been splendidly debunked in two recent articles by historian of science Frank Sulloway. His arguments form the basis of this essay (see bibliography).

The thirteen species of Darwin's finches form a closely knit genealogical group of widely divergent life styles—a classic case of adaptive radiation into a series of roles and niches that would be filled by members of several bird families in more conventional, and crowded, continental situations. We get our major clues about the adaptations of these species from the shapes of their bills. Three species of ground finches have large, medium, and small beaks, while a fourth grows a sharp, pointed bill. All are adapted to eating differing seeds of appropriate size and hardness. Two species feed on cactus and another on mangroves. Four inhabit trees—of these, one is a vegetarian, while the other three eat large, medium, and small insects, respectively. A twelfth species closely resembles warblers in form and habits; while the thirteenth, the most curious of all, uses twigs and cactus spines as tools to extract insects from crevices in tree trunks.

The fine work of the great British ornithologist David Lack has taught us that the thirteen species evolved and became more distinct through a four-stage process of colonization, isolation and speciation, reinvasion, and perfecting of adaptation in competition. Lack also gave the birds their felicitous name of "Darwin's finches," in his 1947 book of the same title. But, contrary to anachronistic legend, this classic description of speciation is not a story that Darwin ever knew.

Darwin visited the Galápagos in September and October 1835, landing on only four of the islands. At sea, sometime during the middle of 1836, he penned a famous statement in his *Ornithological Notes,* a major source for the legend that his Galápagos experiences directly converted him to evolution and that the finches inspired his new view of life:

> When I recollect, the fact from the form of the body, shape of scales and general size, the Spaniards can at once pronounce, from which Island any Tortoise may have been brought. When I see these Islands in sight of each other, and possessed of but a scanty stock of animals, tenanted by these birds, but slightly differing in structure and filling the same place in Nature, I must suspect that they are only varieties. The only fact of a similar kind of which I am aware, is the constant asserted difference—between the wolf-like Fox of East and West Falkland Islds.—If there is the slightest foundation for these remarks the zoology of Archipelagos will be well worth examining; for such facts would undermine the stability of Species.

First of all, the "birds" of this passage are Galápagos mockingbirds, not finches. Darwin did notice that three of the four islands he visited contained distinctly different mockingbirds. At face value, this statement seems to display a strong preference for evolution; it certainly raises the possibility. But a familiarity with nineteenth-century zoological terminology suggests an alternate interpretation. All creationists admitted that species often differentiated into mildly distinct forms in situations, as on island chains and archipelagoes, where populations could become isolated in differing circumstances of ecology and climate. These local races were called varieties, and they did not threaten the created and immutable character of a species' essence. Properly translated from the terminology of his time, Darwin says in this famous statement that either the tortoises and mockingbirds are merely varieties—in which case they do not threaten his creationist views—or they have become separate species, in which case they do. He briefly considered evolution by admitting the second possibility, but he ultimately drew back while still at sea by tentatively deciding (incorrectly, for the mockingbirds at least) that the island forms were only varieties. Darwin's memories as an old man confirm this view that he only briefly flirted with, and then rejected, evolution while on the *Beagle*. He wrote to the German naturalist Otto Zacharias in 1877: "When I was on board the *Beagle* I believed in the permanence of species, but, as far as I can remember, vague doubts occasionally flitted across my mind."

A second statement, taken in conjunction with a misreading of the *Ornithological Notes,* might also be considered as evidence that Darwin became an evolutionist at sea in 1836. He wrote in his pocket journal: "In July opened first notebook on 'Transmutation of Species'—Had been greatly struck from about Month of previous March on character of S. American fossils—and species on Galápagos Archipelago. These facts origin (especially latter) of all my views." We know that he started the first Transmutation notebook in July 1837, and we might therefore interpret the "previous March" as 1836, about the time that he penned the *Ornithological Notes* at sea. But the previous March might as well be 1837 when, as we shall soon see, he was in London learning from specialists at the Zoological Society about the true character of his Galápagos collections—a set of phenomena that he had failed to observe during his own visit.

What, then, did Darwin see on the Galápagos, and what did he miss? Three groups of animals have come down through history as the most famous evolutionary laboratories of the Galápagos: mockingbirds, tortoises, and finches. Only for the mockingbirds did Darwin make the key observation that underlies the evolutionary tale developed later (although, as we have seen, Darwin first explicitly rejected the evolutionary reading for a different interpretation). In short, he noticed that varying forms (later recognized as true species, although Darwin originally labeled them varieties) inhabited the different islands he visited. He landed first at Chatham Island, then at Charles, and he realized that he could distinguish the Charles Island mockingbird from the form he had previously collected at Chatham. Thus, he collected more mockingbirds wherever he landed and he carefully kept the separate island collections well labeled and distinct. He could not distinguish the Albemarle mockingbird, on the third island he visited, from the Chatham form, but the James Island bird represented a third, distinct variety (by his original interpretation).

Galápagos tortoises are all of one species, but most islands feature their own recognizable subspecies. These span an impressive range of form, from smooth, dome-shaped carapaces to the peculiar saddlebacks, with a pronounced hump in the carapace just above the head. Darwin missed this story completely. He never even noted the saddlebacks. Moreover, his basic concept of these tortoises virtually guaranteed that he would not be able to make the key observation.

Nicholas Lawson, the vice-governor, told Darwin that "the tortoises differed from the different islands, and that he could with certainty tell from which island any one was brought" (although distinctions abound, this statement is overly optimistic and modern experts cannot always distinguish each island). But Darwin, by his own admission, made little of this information, writing in the 1845 edition of the *Beagle Voyage:*

> I did not for some time pay sufficient attention to this statement, and I had already partially mingled together the collections from two of the islands. I never dreamed that islands, about fifty or sixty miles apart,

and most of them in sight of each other, formed of precisely the same rocks, placed under quite similar climate, rising to a nearly equal height, would have been differently tenanted.

As the result of a common error in classification, Darwin was ill-disposed to consider the differences between islands as evolutionarily (or even taxonomically) meaningful. Darwin accepted the general view that Galápagos tortoises were not taxonomically distinct but belonged to the species *Testudo indicus*, the giant land tortoise of the Aldabra Islands in the Indian Ocean. They had only recently been brought, so the false story continued, to the Galápagos by buccaneers. Hence, differences among islands, if they existed at all, could only be immediate and superficial—inspired by harsh climates at the time of introduction. Moreover, Darwin never saw live saddleback tortoises. He only observed living tortoises on James and Chatham islands, and both contain nearly indistinguishable versions of the dome-shaped carapace.

Still, Darwin cannot be entirely excused from a charge of some carelessness in observation. He did have an opportunity to observe the saddleback form but either failed to do so or recorded no impression. The Charles Island race was extinct when Darwin landed, but old carapaces were abundant at the settlement, where they served as flowerpots. Moreover, Darwin showed singularly little interest in preserving specimens for comparison among islands, a sure sign that he did not regard Lawson's statement as significant (much to his later regret). Captain Fitzroy took thirty large Chatham tortoises on board to beef up the *Beagle*'s supply of fresh meat during the long Pacific crossing. Sulloway remarks:

But Darwin and the other crew members gradually ate their way through the evidence that eventually, in the form of hearsay, was to revolutionize the biological sciences. Regrettably, not one of the thirty Chatham

Island carapaces reached England, having all been thrown overboard with the other inedible remains.

Darwin's reaction to the Galápagos finches included even more error and misunderstanding. Again, he showed no appreciation for the importance of differences between islands. In fact, he didn't even bother to record or label the islands that had housed his specimens. Only three of his thirty-one finches are identified by island in the *Ornithological Notes,* all members of a highly distinctive species that Darwin remembered seeing only on James Island. He later wrote with regret in the *Voyage of the Beagle:* "Unfortunately most of the specimens of the finch tribe were mingled together." Secondly, he failed completely to collect any finches on one of the islands he visited—Albemarle. True, he was there for only part of a day, but his own diary records an abundance of easily collectable finches at a spring near Bank's Cove: "To our disappointment the little pits in the Sandstone contained scarcely a gallon of water and that not good. It was however sufficient to draw together all the little birds in the country; Doves and Finches swarmed around its margin."

Third, except for cactus and warbler finches, Darwin failed to observe any distinction in diet among the species and believed erroneously that they all ate the same kinds of food. Thus, he could not have reconstructed our modern story, even if he had been inclined to evolutionary views.

Fourth, Darwin's entire style of collection on the Galápagos strongly reflected his creationist presuppositions. Evolutionists see variation as fundamental, as the raw material of evolutionary change. Species can only be well defined by collecting many specimens and defining the spectrum of their variation. Creationists believe that each species is endowed with a fixed essence. Variation is a mere nuisance, a confusing array of environmentally induced departures from an ideal form. Creationists tend to gather a limited number of specimens from each species and to concentrate on procuring individuals closest to the essential form. Darwin collected very few specimens, generally only a male and female of each species. In all, he procured but thirty-one finches from the Galápagos. By contrast, a 1905–1906 California Academy of Sciences expedition, dispatched to study evolution explicitly, brought back more than 8,000 specimens.

Fifth, and most importantly, the finches tell no evolutionary tale unless you recognize that, despite their outward differences in form and behavior, all form a tightly knit genealogical group. But Darwin, while on the Galápagos, was fooled by the stunning diversity and failed to recognize Darwin's finches as a taxonomic entity. He referred the cactus finch to a family of birds that includes orioles and meadowlarks, and he misclassified the warbler finch as either a wren or warbler. Those that he recognized as finches, he divided into two distantly related groups within the family. Sulloway remarks: "As for Darwin's supposed insight into evolution by adaptive radiation while he was still in the Galápagos, the more the various species of finch exhibited this remarkable phenomenon, the more Darwin mistook them at the time for the forms they were mimicking."

The theoretical source of Darwin's error lies in a fairly arcane principle of the creationist style of taxonomy that he followed. If animals are created according to a rational and general plan in the Deity's mind, then certain "key" characters might be clues to taxonomic structure at different levels. For example, variation in such "superficial" characters as size and shape might define different species, while variation in such "fundamental" traits as the form of essential organs might record the more important differences between genera and families. Ideally, a hierarchy of key characters should define taxonomic levels. Darwin tried to follow such a system in his preliminary *Beagle* classifications. Species within a bird genus should differ in plumage, while genera should be separated by such characters as the form of the beak. Darwin's finches are all similar in plumage, but differ greatly in styles of feeding and, consequently, in the shapes of their beaks. By Darwin's creationist hierarchy of key characters, they belonged to different genera or families.

The key character hierarchy makes no sense in an evolutionary context. Characters that define genera in one situation might vary widely among species within another group. Bills may define feeding types, and feeding types may usually distinguish genera on continents. But if only one kind of small bird manages to reach an oceanic archipelago and then diversifies, in the absence of competitors, into a wide range of niches and feeding types, then the traditional criterion for genera—form of the bill—will now differ among closely related species. In the blooming and buzzing confusion of evolution, as opposed to the order of a creator's mind, responses to local preset environments, not rules of change, determine what parts of the body will be modified in any particular case. Behavior and plumage in one place; feeding and shape of the beak in another. There is no such thing as an invariably "specific" or "generic" character.

In summary, then, Darwin entered and left the Galápagos as a creationist, and his style of collection throughout the visit reflected his theoretical stance. Several months later, compiling his notes at sea during the long hours of a Pacific crossing, he briefly flirted with evolution while thinking about tortoises and mockingbirds, not finches. But he rejected this heresy and docked in England, October 2, 1836, as a creationist harboring nascent doubts.

This retelling of the finch story is particularly satisfying because the new version squares so much better than the old legend with Darwin's use of the Galápagos finches throughout his later writing. He never mentioned them in any of the four *Transmutation Notebooks,* which he kept from 1837 to 1839 and which serve as a foundation for his later work. They receive only passing notice in the first (1839) edition of the *Voyage of the Beagle.* To be sure, the second edition (1845) does contain this prophetic statement, written after Darwin had learned that the finches form a closely knit genealogical group.

> Seeing this gradation and diversity of structure in one small, intimately related group of birds, one might really fancy that from an original paucity of birds in

this archipelago, one species had been taken and modified for different ends.

But if the finches made such a belated impression, the impact didn't seem to last. Darwin's finches are not mentioned at all in the *Origin of Species* (1859); the ornithological star of that great book is the domesticated pigeon. Sulloway concludes, rightly I think:

> Contrary to the legend, Darwin's finches do not appear to have inspired his earliest theoretical views on evolution, even after he finally became an evolutionist in 1837; rather it was his evolutionary views that allowed him, retrospectively, to understand the complex case of the finches.

Darwin returned to England in 1836 as an ambitious young man, anxious to make his mark in science; his later, courtly modesty as an old man should not be allowed to mask this youthful vigor. He knew that the key to his reputation lay in the valuable specimens he had collected on the *Beagle*, and therefore he made determined and successful efforts to farm them out to the best specialists and to procure funds for publication of the results. In March 1837 he moved to London, largely to be near the various experts studying his specimens. He began a series of meetings with these men, finally learned the true character of his material, and emerged within a month or two as an evolutionist.

He wrote, in the famous entry in his pocket journal cited earlier, that the character of South American fossils and species of the Galápagos had been the primary catalysts of his evolutionary conversion. Richard Owen, Britain's most eminent vertebrate paleontologist, had agreed to study the fossils and informed Darwin that they represented different, usually larger versions of distinctive animals that still inhabit South America. Darwin recognized that the best interpretation of this "law of succession" cast the ancient forms as evolutionary ancestors of altered modern animals.

The famous ornithologist John Gould (no relation) had

taken charge of the *Beagle*'s birds. Darwin met with him toward the middle of March and learned that the three forms of mockingbirds were separate species, not mere and superficial varieties of a single, created type. Darwin had already proclaimed that such a conclusion (which he had previously rejected) "would undermine the stability of species." Moreover, Gould informed him that twenty-five of his twenty-six Galápagos land birds were new species, but clearly allied to related forms on the South American mainland. Darwin integrated this spatial information with the temporal data that Owen had supplied, and he tilted further toward evolution. The distinct Galápagos birds must be evolutionary descendants of mainland colonists from South America. Darwin was now fully primed for an evolutionary reading of the finches, and Gould's correction of Darwin's errors furnished this piece of the puzzle as well (although Gould himself did not adopt evolutionary views).

Although a creationist in taxonomy, Gould recognized right away that bills could not be used as a key character to separate genera of Galápagos finches. He understood that these birds were not, as Darwin had thought, a heterogeneous assemblage of divergent finches with an unrelated warbler and oriole thrown in, but a peculiar group of thirteen closely related species, which he placed in a single genus with three subgenera. "The bill appears to form only a secondary character," Gould proclaimed. Darwin finally had the basis of an evolutionary story.

Darwin was exhilarated as he converted to evolution and prepared to reread his entire voyage in this new light. But he was also acutely embarrassed because he now realized that his failure to separate finches by islands, no particular problem in a creationist context, had been a serious and lamentable lapse. He couldn't do much with his own collection, beyond probing a faulty and fading memory; but fortunately, three of his shipmates had also collected finches—and since they (ironically) had not collected with an actively creationist theory in mind (with its implied irrelevancy for precise geographic data), they had recorded the islands of collection. As a further irony, one of these collections had been made by Captain Fitzroy himself, later Darwin's implacable foe and the man who stalked around the British Association meeting where Huxley demolished Wilberforce, holding a Bible above his head and exclaiming, "the Book, the Book." (Fitzroy's collection included twenty-one finches, all labeled by island. Darwin also had access to the smaller collections of his servant Syms Covington and of Harry Fuller, who had spent a week collecting with him on James Island.)

Darwin therefore tried to reconstruct the localities of his own specimens by comparing them with the accurately labeled collections of his shipmates and, unfortunately as it turned out, by assuming that the finch story would resemble the pattern of the mockingbirds—with each species confined to a definite island. But since most of the finch species inhabit several islands, this procedure produced a large number of errors. Sulloway reports that substantial doubt still exists about the accuracy of geographic information for eight of fifteen among Darwin's "type" (or name bearing) specimens of finches. No wonder he was never able to make a clear and coherent story of Darwin's finches. No wonder, perhaps, that they never even appeared in the *Origin of Species*.

Why, in conclusion, is this correction of the finch legend of any great importance? Are the two stories really all that different? Darwin, in either case, was greatly influenced by evidence from the Galápagos. In the first, and false, version he understands it all by himself while on the visit. In the second, modified account he requires a nudge (and some substantial corrections) from his friends when he returns to London.

I find a world of difference between the tales for what they imply about the nature of creativity. The first (false) version upholds the romantic and empirical view that genius attains its status from an ability to see nature through eyes unclouded by the prejudices of surrounding culture and philosophical presupposition. The vision of such pure and unsullied brilliance has nurtured most legends in the history of science and purveys seriously false views about the process of scientific thought. Human beings cannot escape their presuppositions and see "purely"; Darwin functioned as an active creationist all through the *Beagle* voyage. Creativity is not an escape from culture but a unique use of its opportunities combined with a clever end run around its constraints. Scientific accomplishment is also a communal activity, not a hermit's achievement. Where would Darwin have been in 1837 without Gould, Owen, and the active scientific life of London and Cambridge?

Once we abandon the alluring, but fallacious, image of Darwin winning his intellectual battle utterly alone at sea, we can ask the really interesting question that begins to probe Darwin's particular genius. Gould was the expert. Gould resolved the details correctly. Gould, a staunch creationist in taxonomy, nonetheless recognized that he had to abandon beaks as key characters. Darwin could accomplish none of this. But Darwin, not Gould, recognized that all the pieces required a stunningly new explanation—evolution—to make a coherent story. The amateur triumphed when the stakes were highest, while the professional got the details right and missed the organizing theme.

Darwin continued to work in this way throughout his career. Somehow, as an amateur, he could cut through older patterns of thought to glimpse new modes of explanation that might better fit an emerging, detailed story constructed by experts who, somehow, could not take the big and final step. But Darwin worked with his culture and with his colleagues. Science is a collective endeavor, but some individuals operate with an enlarged vision—and we would like to know how and why. We can ask no harder question, and I propose no general solution. But we do need to clear away heroic legends before we can begin.

24 | A Short Way to Corn

SINCE IT WAS only a few miles to Tipperary, not the long way of song and legend, I took a detour to visit the town. Soon I felt like the city slicker of that old New England joke. Looking for a small town, he stops before a general store and asks an old-timer, "Where is Pleasantville?" "Don't you move a god-damned inch," comes the reply.

Tipperary, made large by its fame and my imagination, is but one main street with a few stores and houses. This eerie scene repeated itself again and again during my visit to this most beautiful of European lands. For Ireland, contrary to the trend of most other countries, is a depopulated nation. Its current count of some 3 million inhabitants includes but half of the 1840 total. Abandoned homes, farms, and even towns lie strewn about the countryside.

The beginning of the great emigration that so enriched my native city of New York and my current Boston home dates to the great potato famines of 1845 and 1846, when half a million people starved to death and another million left. The potato is a remarkable food. It contains so well balanced an array of nutrients that people can live on virtually nothing else for years on end. Monotonous perhaps, spuds being spuds, but quite viable. Irish peasants often ate nothing but potatoes through the long winter months. But disease attacked the crop in 1845 and virtually destroyed it, producing unprecedented starvation and the great exodus to Liverpool and beyond.

The Irish potato blight illustrates a classic dilemma in agriculture. To produce the "best" plant for maximal yields, farmers and scientists will hone and select for many generations until they obtain just the right combination of features. They will then propagate their entire crop from this improved form. These plants, as offspring of a single parental type, are genetically uniform and depleted in variability. In other words, we trade genetic diversity for an unvarying optimum.

All may be well for a while, but uniform stocks are exquisitely susceptible to the ravages of disease. If some virus, bacterium, or fungus successfully attacks the plants, it can destroy every one, thus devastating the crop. In natural populations, on the other hand, genetic variation among individuals insures that some will enjoy protection against the agent of disease and part of the crop will survive. Since next year's plants are offspring of these immune survivors, populations with abundant variability maintain a natural mechanism to purge themselves of disease.

The Irish, growing their potatoes from a uniform stock, lost their entire crop in 1845. The same story can be told for most agricultural mainstays. Some scholars believe that the mysterious collapse of classic Maya civilization was precipitated by a virus, borne by leaf hoppers dispersed on high-altitude air currents, that wiped out their corn crop virtually overnight. Corn continues to plague us with similar problems. During the summer of 1970, a new mutant strain of Southern Leaf Blight Fungus swept across American cornfields at rates of fifty miles or more a day, devastating all plants bred to contain a genetic element called Texas cytoplasmic male sterility factor.

To avoid this dilemma, breeders try to beef up genetic variability by hybridizing their successful but uniform stocks with different strains. For corn, a major source of potential hybridization lies in a plant of markedly different appearance, the New World grass known as teosinte. For example,

Zea diploperennis, a recently discovered species of teosinte, is the only known source of immunity to three of the major viruses that afflict domestic corn. (This species is also a perennial rather than an annual like corn, thus giving potential substance to an old dream that, by hybridization, breeders might produce a perennial corn that survives from season to season and need not be replanted from seed each year.)

It may seem strange at first that a plant so different in appearance from corn should be sufficiently similar in genetic structure to permit hybridization. True, young plants of corn and teosinte are indistinguishable, but after they flower, the differences in adult structures could hardly be more profound. The business end of corn is a large cob bearing numerous rows of kernels (the technical term, polystichous—simply meaning many rowed—has a lovely ring). The cob and kernels are female, and they reside at the *terminal* end of stout branches *lateral* to the main stem (mark this well, for these positions become crucial in my developing argument). Many people don't recognize this position because corn ears just seem to be stuck to the sides of the main stem. But the husks that so completely enclose the ear are actually remnants of leaves that originally formed on a longer lateral branch. They cover the cob, which is, indeed, terminal on a drastically shortened lateral branch. The central stem bears a male tassel, the source of pollen, at its terminal end. Thus, corn grows separate male and female structures: the tassel, terminal on the main stem, is male; the ears, terminal on lateral branches, are female.

Teosinte, on the other hand, grows a central stem and many long lateral branches of comparable length and strength. Each branch ends in a male tassel. The female ears, quite unlike corn, grow laterally, not terminally, from the lateral branches. The teosinte "ear" is also a miserable analog or runt compared with the majestic ear of corn. It contains (depending on the race of teosinte) six to twelve triangular kernels in two rows (technically distichous) telescoped into one because the triangular ends of the opposing kernels interdigitate. The kernels are surrounded by a

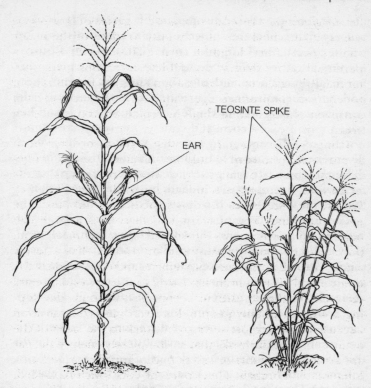

EAR

TEOSINTE SPIKE

In modern corn (left), female ears are terminal on lateral branches; in teosinte (right), male tassels are terminal on lateral branches, while female ears are lateral on the lateral branches. Thus, the modern corn ear is the homolog of a teosinte tassel spike. See text for explanation. REPRINTED FROM NATURAL HISTORY.

stony outer covering and are quite useless as human food unless popped (as in popcorn) or laboriously ground and separated from their inedible covering. (Corn kernels are soft and naked, immediately available for food because their covering structures are not only pliant, but so reduced in size that they surround only the base of the kernel.)

Yet, despite these differences, corn and teosinte hybridize without any impediment, producing cobs of intermediate size. This paradoxical compatibility exists for two basic reasons that reflect the subject of this essay—a disquisition on the ancestry and origin of corn. First, teosinte is probably the direct ancestor of domestic corn (some experts disagree, although no one denies the close relationship). Second, no chromosomal disparities or even simple and consistent differences in single genes have been found between teosinte and corn. (Of course, the two forms could not be so different in appearance without some genetic divergence, but ease of hybridization and our failure to find differences indicate that genetic distinction between the two forms must be minuscule. Indeed, botanists place corn and the annual teosintes in the same species, *Zea mays.*)

The teosinte theory for the origin of corn has always suffered from one major dilemma: How could it happen? How could the teosinte ear, so different from corn, become the modern cob? Corn, like all our major domestic cereals, is a grass. The evolutionary origin of other major grains, wheat for example, presents fewer problems. Wheat ears differ only quantitatively from their wild grassy ancestor— they are essentially just more of the same, but bigger. We can easily understand how agricultural selection could translate a wild ancestor into domestic wheat. But how can you make a corncob from a teosinte ear or from any part of teosinte? They are constructed so differently.

In the standard version of the teosinte hypothesis—which I will reject here in favor of a radical alternative—the teosinte ear is, nonetheless, gradually transformed into a modern corn ear. It adds rows gradually, while the hard outer covering softens and retracts from the kernels. This scenario seems so obvious and so consistent with our usual view of evolutionary transformation. The tradition of gradual change from teosinte ear to corn ear dates at least to Luther Burbank, the great "wizard" of early twentieth-century plant breeding, who claimed that he had transformed teosinte to corn in eighteen generations of selection. He was wrong. He had started, not with teosinte, as he thought, but with a corn-teosinte hybrid—and his selection had merely segregated and accumulated the genetic factors for corn. But his general argument for a gradual transformation of the teosinte ear into a corncob persisted. In a *Scientific American* article of January 1980, George Beadle, one of the great corn scientists of our age, proclaimed that "the cobs can be placed in an evolutionary continuum from teosinte to modern corn on the basis of progressive modifications."

But this theory of gradual derivation from the teosinte ear encounters three great problems, perhaps fatal. First, corn appears suddenly in the archeological record about 7,000 years ago. The earliest ears, to be sure, are not as fat or as many-rowed as a modern cob, but they clearly represent corn, not something in between corn and teosinte. Second, as stated before, breeders have found no consistent genetic difference between corn and teosinte. If corn were the product of long and slow selection from teosinte, a considerable number of genetic changes should have accumulated. Both these arguments are negative and therefore not conclusive. Perhaps sudden appearance merely records our failure to find intermediates; perhaps the absence of genetic difference only means that we haven't looked at the right parts of the right chromosomes.

The third argument is positive and more troubling for the hypothesis that corn ears arose from teosinte ears. Remember the point I asked you to flag some paragraphs back: the positions of teosinte and corn ears are not equivalent. The teosinte ear sprouts laterally from lateral branches; the corn ear grows terminally on lateral branches. In teosinte, the terminal structure on the main lateral branches is a male tassel, not a female ear. Therefore, by position—and I shall say in a moment why position is so important a criterion— the modern female ear of corn is equivalent to (or, as we say in technical parlance, is the homolog of) a male tassel spike.

This homology of male tassel spike to female ear has long been recognized (and stated) by many corn experts, but no one has previously exploited this fact to develop a hypothesis for the origin of corn. The obvious theory suggested by this homology may, at first, sound absurd, but it solves plausibly and with elegance all the classical problems of the teosinte hypothesis. In short, this new theory proposes that corn ears evolved rapidly from male tassel spikes by shortening of the lateral branches and suppression of teosinte ears below. Instead of a slow and continuous enlargement of female teosinte ears, we envision an abrupt transformation of male tassel spikes to small and primitive versions of a modern female corn ear.

Hugh H. Iltis, professor of botany and director of the Herbarium at the University of Wisconsin in Madison, developed this heterodox theory and recently published it in America's leading professional journal (see bibliography).* I have no corn credentials and cannot make any proclamation about the truth or falsity of this intriguing idea. But I do want to illustrate its status as a plausible, potential example of an evolutionary process often dismissed with ridicule for want of understanding—the so-called hopeful monster.

We call parts of two organisms "homologous" when they represent the same structure by a criterion of evolutionary descent from a common ancestor. No concept is more important in unraveling the pathways of evolution, for homologies record genealogy, and false conclusions about homology invariably lead to incorrect evolutionary trees.

Homologous structures need not look alike. Indeed, the standard examples invoke organs quite dissimilar in form and function, for these "classics" are chosen to illustrate the idea that mere resemblance does not qualify as a criterion. Examples include the homology of the hammer and anvil bones of the mammalian middle ear with the jaw articulation bones of reptiles; and the lung of land vertebrates with the air bladder of bony fishes.

How then can we recognize homology and thereby recon-

---

*Iltis's unconventional theory quickly unleashed the expected volley of criticism from defenders of the more traditional views. Readers wishing to pursue the controversy further might begin with the critiques of two "grand old men" of corn studies (Walton C. Galinat and Paul C. Mangelsdorf) and Iltis's response, all published in Science, September 14, 1984, pp. 1093–1096, soon after this essay originally appeared. Mangelsdorf's book Corn (Harvard University Press, 1974) contains a wealth of detail on our hemisphere's greatest contribution to human nutrition.

struct the pathways of evolution? This most difficult question in evolutionary theory has no definite answer. No single criterion works in every case; all rules have well-known exceptions. We must evaluate proposed homologies by all available standards and accept or reject a hypothesis by the joint and independent affirmation of several criteria. Similarity in early embryology often works well for structures that become very different in adults: early mammalian embryos first develop their ear bones at the ends of their jaws—and this fact harmonizes with a well-established fossil sequence showing continual decrease of these two jaw bones and their eventual movement to the middle ear. But truly homologous organs may be modified by evolutionary changes in embryos that mask the pathways of descent.

A seemingly superficial detail—simple spatial relation with other parts—often serves well as a criterion of homology. As the old song goes, the foot bone truly is connected to the ankle bone, and such fundamental relationships are not easily altered in evolution. Thus, the so-called "positional criterion" of homology is probably the most respected and most often utilized of all standards. And by this criterion, modern female corn ears must have descended from teosinte male tassel spikes (for both features are alike in position at the terminal ends of lateral branches), and not from female teosinte ears.

Lest it seem absurd that male structures be transformed into female organs of such different appearance, I remind readers that male and female parts often have a common basis in embryology, one developing directly from the other under the influence of different hormones. The external genitalia of all mammals, for example, begin as female structures: the clitoris enlarges, folds over and fuses to form a cylinder with a central tube, the male penis; the *labia majora* expand and fuse at the midline to form a scrotal sac. In essay 11 of *Hen's Teeth and Horse's Toes,* I used these equivalences to argue that the remarkable male-mimicking genitalia of female spotted hyenas arise automatically from these common pathways of sexual development, because females of this species secrete unusually high levels of testosterone during growth and become both larger than males and dominant over them.

Tassel spike and corn ear are also equivalent structures and the transformation of one to the other is equally plausible. Indeed, such interchanges often occur as teratologies, or abnormalities of development, in modern corn. Male tassel spikes may grow as female ears or partial ears with male ends, for several reasons: genetic mutations and diseases that drastically shorten the central branch, for example. Iltis sent me the accompanying photograph of such a feminized tassel, sold as an ordinary ear of corn at Kohl's supermarket in Madison, Wisconsin, for thirty-nine cents. The central axis has grown as a complete female ear. The three lateral branches are female at the bottom, grading to male at the top, with the male parts "arranged," Iltis tells me, "exactly as in any maize or teosinte tassel branch." Of course, such teratologies only show the interchangeability of corn tassels and corn ears, not the evolutionary derivation of corn ears from teosinte tassels. But they surely illustrate, by strong analogy, why the genealogical path from teosinte tassel to corn ear remains so reasonable, however peculiar at first hearing.

In calling his theory the "catastrophic sexual transmutation," Iltis forcefully identifies its two outstanding and unconventional properties. First, using the positional criterion of homology as a guide, female corn ears arose by transmutation of a male teosinte tassel spike, not by gradual enlargement of a female teosinte ear. Second, the transformation occurred rapidly under the guidance of little (or even no) genetic change, despite the sudden and striking alteration of form. I shall try to epitomize Iltis's argument in the following basic steps:

1. In both corn and teosinte, hormones are distributed along simple gradients in long stems, with male zones at the tops passing through a threshold to female zones below.

2. A gradient in time of differentiation during growth also accompanies this hormonal distribution. Structures at the tops of stems develop earlier than those

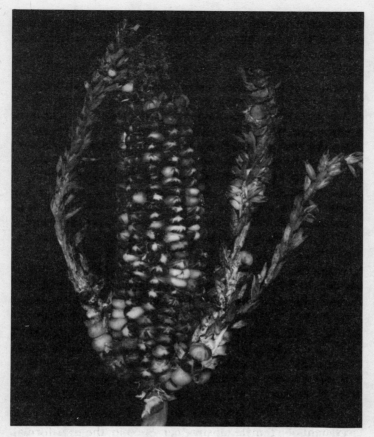

A corn "monster" purchased for thirty-nine cents at a Madison, Wisconsin, supermarket by Hugh Iltis. PHOTO COURTESY OF HUGH ILTIS.

lower down. On a lateral teosinte branch, the terminal male tassel differentiates before the female ears
below.

3. The nutritional needs of a male tassel are small, those of
a female ear (particularly a large and polystichous ear of
corn) much larger. The differentiation of a tassel at the
terminal end of a branch still leaves most nutrients avail-

able for development of subsequent female structures below (see point 2).

4. If a terminal male tassel spike transmutes abruptly to a female ear, this ear would immediately become a sink for all available nutrients and might automatically suppress the development of any subsequent female structure lower down on the branch.

5. The initial step of the catastrophic sexual transmutation, given points 1 through 4 above, might therefore require nothing more than a marked shortening of a teosinte lateral branch. The shortening would move the branch's tip from a masculine to a feminine zone (point 1). The terminal structure would then differentiate first as a female ear (points 1 and 2). This terminal ear would appropriate all nutrients and suppress the development of any structures below, including the usual female ears of teosinte (points 3 and 4).

6. Although the shortening of a branch might induce a profound and varied set of automatic consequences (point 5), the initiating change (the shortening itself) is simple and may require only a trifling genetic alteration, perhaps the mutation of a single gene. The initial change might even require no genetic modification at all, for several corn smuts and viruses, or even a simple environmental change to cooler night temperatures or shorter days, can lead to a feminization of central tassel spikes in corn.

7. Of course, the initial product of such shortening and feminization would not be a full-blown modern corn ear. This first step would probably yield a cob with a few rows of female kernels at the base and male structures above. The production of a polystichous, or many rowed, ear remains problematical. One hypothesis envisions the conjunction of several tassel segments (as the branch shortens) and their subsequent junction and twisting to form the polystichous ear. Remember, though, that the conventional theory of derivation from a two-rowed teosinte ear encounters the very same problem and proposes the same basic resolution. As Iltis points out, the teosinte tassel is a better candidate than the teosinte ear

for such a hypothetical process. The teosinte ear is, as biologists say, a strongly "canalized" structure—one that develops in basically the same way in all individuals of a race, without much variation from plant to plant. It is always two rowed and few kerneled. The tassel spike, on the other hand, is far more variable and prolific in individual units (all suitable for change to kernels). By transforming a male variant with maximal rows and units, the first step might bring us far closer to an ear of corn than any initial change from a teosinte ear could accomplish.

8. The small, initial ear of the catastrophic sexual transmutation is immediately useful as human food. Farmers therefore propagate the kernels and select future generations from plants bearing the largest ears. Ordinary agricultural selection therefore builds the bigger and fuller ear from its initial, rather runty but still useful, condition.

As its major general feature, Iltis's theory proposes that a small genetic change, initiating one basic modification of form (shortening of lateral branches), automatically engenders a major alteration of structure (transformation of male tassel spike to female ear) by "playing off" the inherited sexual and developmental system (hormonal gradients from male to female along a stem, and gradients in differentiation permitting terminal structures to develop first). This theory may therefore serve as a remarkable exemplar of a process long ridiculed by conventional evolutionists but, in my view, eminently plausible in certain cases—the "hopeful monster," a "saltational" view for the origin of novel morphological structures and species (evolution by jumps). The great German geneticist Richard Goldschmidt proposed this idea in a series of works, culminating in his 1940 book, *The Material Basis of Evolution* (recently reprinted by Yale University Press with an introduction by yours truly). Goldschmidt's hopeful monster became the whipping boy of orthodox Darwinians, with their preferences for gradual and continuous change, and his theory suffered the unkindest fate of all—to lie unread and misunderstood while being ridiculed in a caricatured version.

In their caricatured form, hopeful monsters are dismissed for three reasons. Iltis's proposal for the origin of corn beautifully illustrates the theory in its correct and subtle form, as Goldschmidt presented it, and provides a specific antidote for all three arguments. First, detractors claim that Goldschmidt's theory represents a surrender to ignorance, a reliance on some quirky and capricious accident, once in a great while useful by sheer happenstance. Yet don't we know that virtually all major mutations are harmful? Goldschmidt acknowledged that most macromutants are inviable—truly hopeless monsters in his words. The hopeful few won their status precisely because they achieved their abruptly altered form within the constraints imposed by an inherited developmental system. Hopeful monsters are not any old odd change, but large-scale modifications along the established pathways of ordinary sexual and embryological development. In Iltis's theory, inherited hormonal and developmental gradients along a branch permit the sudden transformation of tassel spike to ear. Big changes in harmony with—and produced along—ordinary pathways of development need not be inviable, for they lie within inherited possibilities of basic organization.

Second, hopeful monsters have been rejected because they supposedly propose unknown and large-scale discombobulations of genetic systems. Indeed, late in his career, Goldschmidt did unfortunately confuse his early notion of saltational change in form with a later theory of abrupt and substantial genetic change—the so-called systemic mutation. But, in Goldschmidt's initial version, the hopeful monster arose as a consequence of small—and therefore both plausible and orthodox—genetic changes that produce large effects upon form because they alter early stages of development with cascading effects upon subsequent growth. Iltis proposes a small (or even no) genetic change as a basis for shortening the lateral branches and producing the saltation from tassel spike to ear as an automatic consequence of developmental patterns.

Third, whom shall the hopeful monster choose as mate? It is only an individual, however well endowed, and evolution requires the spreading of favorable traits through populations. The offspring of two such different forms as a normal individual and a hopeful monster will probably be sterile or at least, in their peculiarly hybrid state, no match for normals in natural selection. But Iltis's theory avoids this problem by calling upon human aid to propagate the seeds. The catastrophically transmutated teosinte plant is still a viable creature, with a male tassel on its central spike and female ears in terminal positions on its lateral branches.

Finally, one last interesting and unusual feature of Iltis's theory: it invokes human feedback, not only to improve the initial ear by conventional selection but also to make it a viable structure in the first place—a striking example of interaction between two disparate species in nature. The corn ear, as a natural object, may well be unworkable—for the husks, which firmly enclose the cob as a result of short-ening the lateral branch so drastically, prevent any dispersal of seeds (kernels). In a state of nature, the ear would simply rot where it fell or would seed new plants so close to each other that no offspring would reach full maturity. But farm-ers can peel off the husks and plant the seeds—converting a hopeless to a most hopeful and useful monster.

Corn is the world's third largest crop, not far behind wheat and rice. As the original staple of New World peo-ples, it built the civilizations of an entire hemisphere. Today we grow 270 million acres of corn a year, producing nearly 9 billion bushels. Most of it does not end up in tacos or corn chips, but as animal feed—the primary source for our car-nivorous appetites. We need corn for a comfortable life, but corn needs us as well, simply to survive.

7 | Life Here and Elsewhere

25 | Just in the Middle

THE CASE for organic integrity was stated most forcefully by a poet, not a biologist. In his romantic paean *The Tables Turned,* William Wordsworth wrote:

> Sweet is the lore which nature brings;
> Our meddling intellect
> Misshapes the beauteous form of things:
> We murder to dissect.

The whiff of anti-intellectualism that pervades this poem has always disturbed me, much as I appreciate its defense of nature's unity. For it implies that any attempt at analysis, any striving to understand by breaking a complex system into constituent parts, is not only useless but even immoral.

Yet caricature and dismissal from the other side have been just as intense, if not usually stated with such felicity. Those scientists who study biological systems by breaking them down into ever smaller parts, until they reach the chemistry of molecules, often deride biologists who insist upon treating organisms as irreducible wholes. The two sides of this oversimplified dichotomy even have names, often invoked in a derogatory way by their opponents. The dissectors are "mechanists" who believe that life is nothing more than the physics and chemistry of its component parts. The integrationists are "vitalists" who hold that life and life alone has that "special something," forever beyond

| THE FLAMINGO'S SMILE

the reach of chemistry and physics and even incompatible with "basic" science. In this reading you are, according to your adversaries, either a heartless mechanist or a mystical vitalist.

I have often been amused by our vulgar tendency to take complex issues, with solutions at neither extreme of a continuum of possibilities, and break them into dichotomies, assigning one group to one pole and the other to an opposite end, with no acknowledgment of subtleties and intermediate positions—and nearly always with moral opprobrium attached to opponents. As the wise Private Willis sings in Gilbert and Sullivan's *Iolanthe:*

I often think it's comical
How nature always does contrive
That every boy and every gal
That's born into the world alive
Is either a little Liberal
Or else a little Conservative!
  Fal la la!

The categories have changed today, but we are still either rightists or leftists, advocates of nuclear power or solar heating, pro choice or against the murder of fetuses. We are simply not allowed the subtlety of an intermediate view on intricate issues (although I suspect that the only truly important and complex debate with no possible stance in between is whether you are for or against the designated hitter rule—and I'm agin it).

Thus, the impression persists that biologists are either mechanists or vitalists, either advocates of an ultimate reduction to physics and chemistry (with no appreciation for the integrity of organisms) or supporters of a special force that gives life meaning (and modern mystics who would deny the potential unity of science). For example, a popular article on research at the Marine Biological Laboratory of Woods Hole (in the September-October 1983 issue of *Harvard Magazine*) discusses the work of a scientist with a physicist's approach to neurological problems:

In the parlance of philosophers of science, [he] could be considered a "reductionist" or "mechanist." He believes that fundamental laws of mechanics and electromagnetism suffice to account for all phenomena at this level. Vitalists, in contrast, maintain that some vital principle, some spark of life, separates living from nonliving matter. Thomas Hunt Morgan, a confirmed vitalist, once remarked acidly that scientists who compared living organisms to machines were like "wild Indians who derailed trains and looked for the horses inside the locomotive." Most mechanists, in turn, regard their opponents' vital principle as so much black magic.

But this dichotomy is an absurd caricature of the opinions held by most biologists. Although I have known a few mechanists, as defined in this article, I don't think that I have ever met a vitalist (although the argument did enjoy some popularity during the nineteenth century). The vast majority of biologists, including the great geneticist T. H. Morgan (who was as antivitalist as any scientist of our century), advocate a middle position. The extremes may make good copy, and a convenient (if simplistic) theme for discussion, but they are occupied by few, if any, practicing scientists. If we can understand this middle position, and grasp why it has been so persistently popular, perhaps we can begin to criticize our lamentable tendency to dichotomize complex issues in the first place. I therefore devote this essay to defining and supporting this middle way by showing how a fine American biologist, Ernest Everett Just, developed and defended it in the course of his own biological research.

The middle position holds that life, as a result of its structural and functional complexity, cannot be taken apart into chemical constituents and explained in its entirety by physical and chemical laws working at the molecular level. But the middle way denies just as strenuously that this failure of reductionism records any mystical property of life, any special "spark" that inheres in life alone. Life acquires its own principles from the hierarchical structure of nature.

As levels of complexity mount along the hierarchy of atom, molecule, gene, cell, tissue, organism, and population, new properties arise as results of interactions and interconnections emerging at each new level. A higher level cannot be fully explained by taking it apart into component elements and rendering their properties in the absence of these interactions. Thus, we need new, or "emergent," principles to encompass life's complexity; these principles are additional to, and consistent with, the physics and chemistry of atoms and molecules.

This middle way may be designated "organizational," or "holistic"; it represents the stance adopted by most biologists and even by most physical scientists who have thought hard about biology and directly experienced its complexity. It was, for example, espoused in what may be our century's most famous book on "what is life?"—the short masterpiece of the same title written in 1944 by Erwin Schrödinger, the great quantum physicist who turned to biological problems at the end of his career. Schrödinger wrote:

> From all we have learnt about the structure of living matter, we must be prepared to find it working in a manner that cannot be reduced to the ordinary laws of physics. And that not on the ground that there is any "new force" or what not, directing the behavior of the single atoms within a living organism, but because the construction is different from anything we have yet tested in the physical laboratory.

Schrödinger then presents a striking analogy. Compare the ordinary physicist to an engineer familiar only with the operation of steam engines. When this engineer encounters, for the first time, a more complicated electric motor, he will not assume that it works by intrinsically mysterious laws just because he cannot understand it with the principles appropriate to steam engines: "He will not suspect that an electric motor is driven by a ghost because it is set spinning by the turn of a switch, without boiler and steam."

Ernest Everett Just, a thoughtful embryologist who developed a similar holistic attitude as a direct consequence of his own research, was born 100 years ago in Charleston, South Carolina.* He graduated as valedictorian of Dartmouth in 1907 and did most of his research at the Marine Biological Laboratory of Woods Hole during the 1920s. He continued his work at various European biological laboratories during the 1930s, and was briefly interned by the Nazis when France fell in 1940. Repatriated to the United States, and broken in spirit, he died of pancreatic cancer in 1941 at age fifty-eight.

Just began as an experimentalist, studying problems of fertilization at the cellular level, and in the great tradition of careful, descriptive research so characteristic of the "Woods Hole school." As this work developed, and particularly after he left for Europe, his career entered a new phase: he became fascinated with the biology of cell surfaces. This shift emerged directly from his interest in fertilization and his particular concern with an old problem: How does the sperm penetrate an egg's outer membrane, and how does the egg's surface then react in physical and chemical terms? At the same time, Just's work took on a more philosophical tone (although he never abandoned his experiments), and he slowly developed a holistic, or organizational, perspective midway between the caricatured extremes of classical mechanism and vitalism. Just expounded this biological philosophy, a direct result of his growing concern with the properties of cell surfaces considered as wholes, in *The Biology of the Cell Surface,* published in 1939.

Just's early work on fertilization was a harbinger of things to come. He was not particularly interested in how the genetic material of egg and sperm fuse and then direct the subsequent architecture of development—a classical theme of the reductionist tradition (an attempt to explain the properties of embryology in terms of genes housed in a controlling nucleus). He was more concerned with the effects that fertilization imposes upon the entire cell, partic-

*I wrote this essay in 1983, for the centenary of Just's birth.

ularly its surface, and on the interaction of nucleus and cytoplasm in subsequent cell division and differentiation of the embryo.

Just had an uncanny knack for devising simple and elegant experiments that spoke to the primary theoretical issues of his day. In his very first paper, for example, he showed that, for some species of marine invertebrates at least, the sperm's point of entry determines the plane of first cleavage (the initial division of the fertilized egg into two cells). He also proved that the egg's surface is "equipotential"—that is, the sperm has an equal probability of entering at any point. At this time, biologists were pursuing a vigorous debate (here we go with dichotomies again) between preformationists who held that an embryo's differentiation into specialized parts and organs was already prefigured in the structure of an unfertilized egg, and epigeneticists who argued that differentiation arose during development from an egg initially able to form any subsequent structure from any of its regions.

By showing that the direction of cleavage followed the happenstance of a sperm's penetration (and that a sperm could enter anywhere on the egg's surface), Just supported the epigenetic alternative. This first paper already contains the basis for Just's later and explicit holism—his concern with properties of entire organisms (the egg's complete surface) and with interactions of organism and environment (the epigenetic character of development contrasted with the preformationist view that pathways of later development lie within the egg's structure).

I believe that Just's mature holism had two primary sources in his earlier experimental work on fertilization. First, Just distinguished himself at Woods Hole as the great "green thumb" of his generation. He was a stickler for proper procedure and cleanliness in the laboratory. He had an uncanny rapport with the various species of marine invertebrates that inhabit the waters about Woods Hole. He knew where to find them and he understood their habits intimately. He could extract eggs and keep them normal and healthy under laboratory conditions. He became the chief source of technical advice for hotshot young researchers who had mastered all the latest techniques of experimentation but knew little natural history.

Just therefore understood better than anyone else the importance of healthy normality in eggs used for experiments on fertilization—the integrity of whole cells in their ordinary conditions of life could not be compromised. Over and over again, he showed how many famous experiments by eminent scientists had no validity because they used moribund or abnormal cells and their results could be traced to these "unlifelike" conditions, not to the experimental intervention itself. For example, Just refuted an important set of experiments on abnormalities of development produced when eggs are fertilized by sperm of another species. He proved that the peculiar patterns of embryology must be traced, not to the foreign sperm itself, but to the moribund state of eggs produced by environmental conditions (of temperature and water chemistry) necessary to induce the abnormal fertilization, but uncongenial for the eggs' good health.

Just derided the lack of concern for natural history shown by so many experimenters who knew all the latest about fancy physics and chemistry, but ever so little about organisms. They referred to their eggs and sperm as "material" (I have the same reaction to modern reductionists who call the living cells and organs of their experiments a "preparation") and accepted their experimental objects in any condition because they couldn't distinguish normality from abnormality: "If the condition of the eggs is not taken into account," Just wrote, "the results obtained by the use of sub-normal eggs in experiments may be due wholly or in part to the poor physiological condition of the eggs."

Second, and more important, Just's twenty years of research on fertilization led directly, almost inexorably, to his interest in the cell's surface and to his holistic philosophy. Since his work, as previously mentioned, centered upon the changes that cell surfaces undergo during fertilization, Just soon realized that the cell's surface was no simple, passive boundary, but a complex and essential part of cellular organization:

> The surface-cytoplasm cannot be thought of as inert or apart from the living cell-substance. The ectoplasm [Just's name for the surface material] is more than a barrier to stem the rising tide within the active cell-substance; it is more than a dam against the outside world. It is a living mobile part of the cell.

Later, pursuing a common concern of holistic biology, Just emphasized that the cell surface, as the domain of communication between organism and environment, embodies the theme of interaction—an organizational complexity that cannot be reduced to chemical parts:

> It is keyed to the outside world as no other part of the cell. It stands guard over the peculiar form of the living substance, is buffer against the attacks of the surroundings and the means of communication with it.

Moreover, as his major experimental contribution, Just showed that the cell surface responded to fertilization as a continuous and indivisible entity, even though the sperm only entered at a single point. If the surface has such integrity, and if it regulates so many cellular processes, how can we meaningfully interpret the functions of cells by breaking them apart into molecular components?

> Under the impact of a spermatozoon the egg-surface first gives way and then rebounds; the egg-membrane moves in and out beneath the actively moving spermatozoon for a second or two. Then suddenly the spermatozoon becomes motionless with its tip buried in a slight indentation of the egg-surface, at which point the ectoplasm develops a cloudy appearance. The turbidity spreads from here so that at twenty sec-

onds after insemination—the mixing of eggs and sper-
matozoa—the whole ectoplasm is cloudy. Now like a
flash, beginning at the point of sperm-attachment, a
wave sweeps over the surface of the egg, clearing up
the ectoplasm as it passes.

As his work progressed, Just claimed more and more
importance for the cellular surface, eventually going too
far. He wisely denied the reductionistic premise that all
cellular features are passive products of directing genes in
the nucleus, but his alternative view of ectoplasmic control
over nuclear motions cannot be supported either. Moreover, his argument that the history of life records an increasing dominance of ectoplasm, since nerve cells are most
richly endowed with surface material, and since brain size
increases continually in evolution, reflects the common mis-
conception that evolution inevitably yields progress measured by mental advance as a primary criterion. The following passage may reflect Just's literary skill, but it stands as
confusing metaphor, not enlightening biology:

> Our minds encompass planetary movements, mark
> out geological eras, resolve matter into its constituent
> electrons, because our mentality is the transcendental
> expression of the age-old integration between ecto-
> plasm and non-living world.

Finally, Just's work also suffered because he had the misfortune to pursue his research and publish his book just
before the invention of the electron microscope. The cell
surface is too thin for light microscopy to resolve, and Just
could never fathom its structure. He was forced to work
from inferences based upon transient changes of the cell
surface during fertilization—and he succeeded brilliantly in
the face of these limitations. But within a decade of his
death, much of his painstaking work had been rendered
obsolete.

Thus, Just fell into obscurity partly because he claimed
too much and alienated his colleagues, and partly because

he knew too little as a consequence of limited techniques available to him. Yet the current invisibility of Just's biology seems unfair for two reasons. First, he was basically right about the integrity and importance of cell surfaces. With electron microscopy, we have now resolved the membrane's structure—a complex and fascinating story worth an essay in its own right. Moreover, we accept Just's premise that the surface is no mere passive barrier, but an active and essential component of cellular structure. The most popular college text in biology (Keeton's *Biological Science*) proclaims:

> The cell membrane not only serves as an envelope that gives mechanical strength and shape and some protection to the cell. It is also an active component of the living cell, preventing some substances from entering it and others from leaking out. It regulates the traffic in materials between the precisely ordered interior of the cell and the essentially unfavorable and potentially disruptive outer environment. All substances moving between the cell's environment and the cellular interior in either direction must pass through a membrane barrier.

Second, and more important for this essay, whatever the factual status of Just's views on cell surfaces, he used his ideas to develop a holistic philosophy that represents a sensible middle way between extremes of mechanism and vitalism—a wise philosophy that may continue to guide us today.

We may epitomize Just's holism, and identify it as a genuine solution to the mechanist-vitalist debate, by summarizing its three major premises. First, nothing in biology contradicts the laws of physics and chemistry; any adequate biology must conform with the "basic" sciences. Just began his book with these words:

> Living things have material composition, are made up finally of units, molecules, atoms, and electrons, as surely as any non-living matter. Like all forms in na-

ture they have chemical structure and physical properties, are physico-chemical systems. As such they obey the laws of physics and chemistry. Would one deny this fact, one would thereby deny the possibility of any scientific investigation of living things.

Second, the principles of physics and chemistry are not sufficient to explain complex biological objects because new properties emerge as a result of organization and interaction. These properties can only be understood by the direct study of whole, living systems in their normal state. Just wrote in a 1933 article:

> We have often striven to prove life as wholly mechanistic, starting with the hypothesis that organisms are machines! Thus we overlook the organo-dynamics of protoplasm—its power to organize itself. Living substance is such because it possesses this organization— something more than the sum of its minutest parts. . . . It is . . . the organization of protoplasm, which is its predominant characteristic and which places biology in a category quite apart from physics and chemistry. . . . Nor is it barren vitalism to say that there is something remaining in the behavior of protoplasm which our physico-chemical studies leave unexplained. This "something" is the peculiar organization of protoplasm.

In striking metaphor, Just illustrates the inadequacy of mechanistic studies:

> The living thing disappears and only a mere agglomerate of parts remains. The better this analysis proceeds and the greater its yield, the more completely does life vanish from the investigated living matter. The state of being alive is like a snowflake on a windowpane which disappears under the warm touch of an inquisitive child. . . . Few investigators, nowadays, I think, subscribe to the naïve but seriously meant

comparison once made by an eminent authority in biology, namely that the experimenter on an egg seeks to know its development by wrecking it, as one wrecks a train for understanding its mechanism. . . . The days of experimental embryology as a punitive expedition against the egg, let us hope, have passed.

Third, the insufficiency of physics and chemistry to encompass life records no mystical addition, no contradiction to the basic sciences, but only reflects the hierarchy of natural objects and the principle of emergent properties at higher levels of organization:

The direct analysis of the state of being alive must never go below the order of organization which characterizes life; it must confine itself to the combination of compounds in the life-unit, never descending to single compounds and, therefore, certainly never below these. . . . The physicist aims at the least, the indivisible, particle of matter. The study of the state of being alive is confined to that organization which is peculiar to it.

Finally, I must emphasize once again that Just's arguments are not unique or even unusual. They represent the standard opinion of most practicing biologists and, as such, refute the dichotomous scheme that sees biology as a war between vitalists and mechanists. The middle way is both eminently sensible and popular. I chose Just as an illustration because his career exemplifies how a thoughtful biologist can be driven to such a position by his own investigation of complex phenomena. In addition, Just said it all so well and so forcefully; he qualifies as an exemplar of the middle way under our most venerable criterion—"what oft was thought, but ne'er so well expressed."

This essay should end here. In a world of decency and simple justice it would. But it cannot. E. E. Just struggled all his life for judgment by the intrinsic merit of his biological research alone—something I have tried so uselessly (and posthumously) to grant him here. He never achieved this recognition, never came close, for one intrinsically biological reason that should not matter, but always has in America. E. E. Just was black.

Today, a black valedictorian at a major Ivy League school would be inundated with opportunity. Just secured no mobility at all in 1907. As his biographer, M.I.T. historian of science Kenneth R. Manning, writes: "An educated black had two options, both limited: he could either teach or preach—and only among blacks." (Manning's biography, *Black Apollo of Science: The Life of Ernest Everett Just,* was published in 1983 by Oxford University Press. It is a superbly written and documented book, the finest biography I have read in years. Manning's book is an institutional history of Just's life. It discusses his endless struggle for funding and his complex relationships with institutions of teaching and research, but says relatively little about his biological work per se—a gap that I have tried, in some respects, to fill with this essay.)

So Just went to Howard and remained there all his life. Howard was a prestigious school, but it maintained no graduate program, and crushing demands for teaching and administration left Just neither time nor opportunity for the research career that he so ardently desired. But Just would not be beaten. By assiduous and tireless self-promotion, he sought support from every philanthropy and fund that might sponsor a black biologist—and he succeeded relatively well. He garnered enough support to spend long summers at Woods Hole, and managed to publish more than seventy papers and two books in what could never be more than a part-time research career studded with innumerable obstacles, both overt and psychological.

But eventually, the explicit racism of his detractors and, even worse, the persistent paternalism of his supporters, wore Just down. He dared not even hope for a permanent job at any white institution that might foster research, and the accumulation of slights and slurs at Woods Hole eventually made life intolerable for a proud man like Just. If he had fit the mold of an acceptable black scientist, he might have

390 | THE FLAMINGO'S SMILE

survived in the hypocritical world of white liberalism in his time. A man like George Washington Carver, who upheld Booker T. Washington's doctrine of slow and humble self-help for blacks, who dressed in his agricultural work clothes, and who spent his life in the practical task of helping black farmers find more uses for peanuts, was paraded as a paragon of proper black science. But Just preferred fancy suits, good wines, classical music, and women of all colors. He wished to pursue theoretical research at the highest levels of abstraction, and he succeeded with distinction. If his work disagreed with the theories of eminent white scientists, he said so, and with force (although his general demeanor tended toward modesty).

The one thing that Just so desperately wanted above all else—to be judged on the merit of his research alone—he could never have. His strongest supporters treated him with what, in retrospect, can only be labeled a crushing paternalism. Forget your research, deemphasize it, go slower, they all said. Go back to Howard and be a "model for your race"; give up personal goals and devote your life to training black doctors. Would such an issue ever have arisen for a white man of Just's evident talent?

Eventually, like many other black intellectuals, Just exiled himself to Europe. There, in the 1930s he finally found what he had sought—simple acceptance for his excellence as a scientist. But his joy and productivity were short-lived, as the specter of Nazism soon turned to reality and sent him back home to Howard and an early death.

Just was a brilliant man, and his life embodied strong elements of tragedy, but we must not depict him as a cardboard hero. He was far too fascinating, complex, and ambiguous a man for such simplistic misconstruction. Deeply conservative and more than a little elitist in character, Just never identified his suffering with the lot of blacks in general and considered each rebuff as a personal slight. His anger became so deep, and his joy at European acceptance so great, that he completely misunderstood Italian politics of the 1930s and became a supporter of Mussolini. He even sought research funds directly from Il Duce.

Yet how can we dare to judge a man so thwarted in the land of his birth? Yes, Just fared far better than most blacks. He had a good job and reasonable economic security. But, truly, we do not live by bread alone. Just was robbed of an intellectual's birthright—the desire to be taken seriously for his ideas and accomplishments. I know, in the most direct and personal way, the joy and the need for research. No fire burns more deeply within me, and no scientist of merit and accomplishment feels any differently. (One of my most eminent colleagues once told me that he regarded research as the greatest joy of all, for it was like continual orgasm.) Just's suffering may have been subtle compared with the brutalization of so many black lives in America, but it was deep, pervasive, and soul destroying. The man who understood holism so well in biology was not allowed to live a complete life. We may at least mark his centenary by considering the ideas that he struggled to develop and presented so well.

26 | Mind and Supermind

HARRY HOUDINI used his consummate skill as a conjurer to unmask legions of lesser magicians who masqueraded as psychics with direct access to an independent world of pure spirit. His two books, *Miracle Mongers and Their Methods* (1920) and *A Magician Among the Spirits* (1924), might have helped Arthur Conan Doyle had this uncritical devotee of spiritualism been as inclined to skepticism and dedicated to rationalism as his literary creation Sherlock Holmes. But Houdini campaigned a generation too late to aid the trusting intellectuals who had succumbed to a previous wave of late Victorian spiritualism—a distinguished crew, including the philosopher Henry Sidgwick and Alfred Russel Wallace, Charles Darwin's partner in the discovery of natural selection.

Wallace (1823–1913) never lost his interest in natural history, but he devoted most of his later life to a series of causes that seem cranky (or at least idiosyncratic) today, although in his own mind they formed a curious pattern of common thread—campaigns against vaccination, for spiritualism, and an impassioned attempt to prove that, even though mind pervades the cosmos, our own earth houses the universe's only experiment in physical objects with consciousness. We are truly alone in body, however united in mind, proclaimed this first prominent exobiologist among evolutionists (see Wallace's book *Man's Place in the Uni-*

verse: A Study of the Results of Scientific Research in Relation to the Unity or Plurality of Worlds, 1903).

Wallace's basic argument for a universe pervaded by mind is simple. I also regard it as both patently ill-founded and quaint in its failure to avoid that age-old pitfall of Western intellectual life—the representation of raw hope gussied up as rationalized reality. In short (the details come later) Wallace examined the physical structure of the earth, solar system, and universe and concluded that if any part had been built ever so slightly differently, conscious life could not have arisen. Therefore, intelligence must have designed the universe, at least in part that it might generate life. Wallace concluded:

> In order to produce a world that should be precisely adapted in every detail for the orderly development of organic life culminating in man, such a vast and complex universe as that which we know exists around us, may have been absolutely required.

How could a man doubt that his favorite medium might contact the spirit of dear departed Uncle George when evidence of disembodied mind lay in the structure of the universe itself?

Wallace's argument had its peculiarities, but one aspect of his story strikes me as even more odd. During the last decade, like the cats and bad pennies of our proverbs, Wallace's argument has returned in new dress. Some physicists have touted it as something fresh and new—an escape from the somber mechanism of conventional science and a reassertion of ancient truths and suspicions about spiritual force and its rightful place in our universe. To me it is the same bad argument, only this time shorn of Wallace's subtlety and recognition of alternative interpretations.

Others have called it the "anthropic principle," the idea that intelligent life lies foreshadowed in the laws of nature and the structure of the universe. Borrowing the term from an opponent who used it for scorn, physicist Freeman

Dyson proudly labels it "animism," not because the idea is lively or organic but from the Latin *anima*, or "soul." (Dyson's essay, "The Argument from Design," in his fine autobiography, *Disturbing the Universe*, provides a good statement of the argument.)

Dyson begins with the usual profession of hope:

> I do not feel like an alien in this universe. The more I examine the universe and study the details of its architecture, the more evidence I find that the universe in some sense must have known that we were coming.

His defense is little more than a list of physical laws that would preclude intelligent life, were their constants just a bit different, and physical conditions that would destroy or debar us if they changed even slightly. These are, he writes, the "numerical accidents that seem to conspire to make the universe habitable."

Consider, he states, the force that holds atomic nuclei together. It is just strong enough to overcome the electrical repulsion among positive charges (protons), thus keeping the nucleus intact. But this force, were it just a bit stronger, would bring pairs of hydrogen nuclei (protons) together into bound systems that would be called "diprotons" if they existed. "The evolution of life," Dyson reminds us, probably "requires a star like the sun, supplying energy at a constant rate for billions of years." If nuclear forces were weaker, hydrogen would not burn at all, and no heavy elements would exist. If they were strong enough to form diprotons, then nearly all potential hydrogen would exist in this form, leaving too little to form stars that could endure for billions of years by slowly burning hydrogen in their cores. Since planetary life as we know it requires a central sun that can burn steadily for billions of years, "then the strength of nuclear forces had to lie within a rather narrow range to make life possible."

Dyson then moves to another example, this time from the state of the material universe, rather than the nature of its physical laws. Our universe is built on a scale that provides, in typical galaxies like our Milky Way, an average distance between stars of some 20 million million miles. Suppose, Dyson argues, the average distance were ten times less. At this reduced density, it becomes overwhelmingly probable that at least once during life's 3.5-billion-year tenure on earth, another star would have passed sufficiently close to our sun to pull the earth from its orbit, thus destroying all life.

Dyson then draws the invalid conclusion that forms the basis for animism, or the anthropic principle:

> The peculiar harmony between the structure of the universe and the needs of life and intelligence is a manifestation of the importance of mind in the scheme of things.

The central fallacy of this newly touted but historically moth-eaten argument lies in the nature of history itself. Any complex historical outcome—intelligent life on earth, for example—represents a summation of improbabilities and becomes thereby absurdly unlikely. But something has to happen, even if any particular "something" must stun us by its improbability. We could look at any outcome and say, "Ain't it amazing. If the laws of nature had been set up just a tad differently, we wouldn't have this kind of universe at all."

Does this kind of improbability permit us to conclude anything at all about that mystery of mysteries, the ultimate origin of things? Suppose the universe were made of little more than diprotons? Would that be bad, irrational, or unworthy of spirit that moves in many ways its wonders to perform? Could we conclude that some kind of God looked like or merely loved bounded hydrogen nuclei or that no God or mentality existed at all? Likewise, does the existence of intelligent life in our universe demand some preexisting mind just because another cosmos would have yielded a different outcome? If disembodied mind does exist (and I'll be damned if I know any source of scientific evidence for or against such an idea), must it prefer a universe that will generate our earth's style of life, rather than a cosmos filled with diprotons? What can we say against diprotons as markers of preexisting intelligence except that such a universe would lack any chroniclers among its physical objects? Must all conceivable intelligence possess an uncontrollable desire to incarnate itself eventually in the universe of its choice?

If we return now to Wallace's earlier formulation of the anthropic principle, we can understand even better why its roots lie in hope, not impelling reason. First, we must mention the one outstanding difference between Dyson's and Wallace's visions. Dyson has no objection to the prospect of intelligence on numerous worlds of a vast universe. Wallace upheld human uniqueness and therefore advocated a limited universe contained within the Milky Way galaxy and an earth impeccably designed, through a series of events sufficiently numerous and complex to preclude repetition elsewhere, for supporting the evolution of intelligent life. I do not know the deeper roots of Wallace's belief, and I have little sympathy for psychobiography, but the following passage from his conclusion to *Man's Place in the Universe* surely records a personal necessity surpassing simple inference from scientific fact. The preexisting, transcendent mind of the universe, Wallace writes, would allow only one incarnation of intelligence, for a plurality

. . . would introduce monotony into a universe whose grand character and teaching is endless diversity. It would imply that to produce the living soul in the marvellous and glorious body of man—man with his faculties, his aspirations, his powers for good and evil —that this was an easy matter which could be brought about anywhere, in any world. It would imply that man is an animal and nothing more, is of no importance in the universe, needed no great preparations for his advent, only, perhaps, a second-rate demon, and a third or fourth-rate earth.

This major difference in opinion about the frequency of intelligent life should not mask the underlying identity of the primary argument advanced by Wallace and by modern supporters of the anthropic principle: intelligent life, be it rare or common, could not have evolved in a physical universe constructed even a tiny bit differently; therefore, preexisting intelligence must have designed the cosmos. Wallace's description of his supporters could well include Dyson: "They hold that the marvellous complexity of forces which appear to control matter, if not actually to constitute it, are and must be mind-products."

Yet the universe used by Wallace to uphold the anthropic principle could not be more radically different from Dyson's. If the same argument can be applied to such different arrangements of matter, may we not legitimately suspect that emotional appeal, rather than a supposed basis in fact or logic, explains its curious persistence? Dyson's universe is the one now familiar to us all—awesome in extent and populated by galaxies as numerous as sand grains on a sweeping beach. Wallace's cosmos was a transient product of what his contemporaries proudly labeled the "New Astronomy," the first, and ultimately faulty, inferences made from a spectrographic examination of stars.

In Wallace's limited universe, the Milky Way galaxy spans some 3,600 light-years in a cosmos that, by Lord Kelvin's calculation, could not be more than twice as large in total diameter (space beyond the Milky Way would be populated by few, if any, stars). A small "solar cluster" of stars sits in the center of the universe; our own sun lies at or near its outer limit. A nearly empty region extends beyond the solar cluster, followed, at a radius of some 300 light-years from the center, by an inner ring of stars and other cosmic objects. Another and much larger region of thinly populated space lies beyond the inner ring, followed by a much larger, densely filled outer ring, the Milky Way proper, with a span of 600 light-years, and lying 1,200 to 1,800 light-years from the center.

Wallace's version of the anthropic principle holds that life

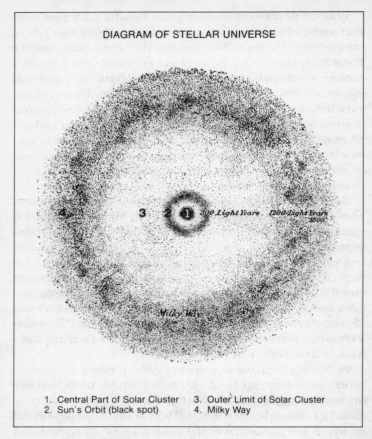

DIAGRAM OF STELLAR UNIVERSE

900 Light Years   1200 Light Years
1800

Milky Way

1. Central Part of Solar Cluster
2. Sun's Orbit (black spot)
3. Outer Limit of Solar Cluster
4. Milky Way

A.R. Wallace's universe, constructed with precision to make human life possible. See text for explanation. FROM WALLACE, 1903. REPRINTED FROM NATURAL HISTORY.

requires each part of this intricate physical universe, and that life could only arise around a sun situated where ours resides by good fortune, at the outer edge of the central solar cluster. All these rings, clusters, and empty spaces must therefore reflect the plan of preexisting intelligence.

Wallace's argument requires that distant stars have a direct and sustaining influence upon our earth's capacity to support life. He flirts with the idea that stellar rays may be good for plants as he desperately tries to argue around a contemporary calculation that the bright star Vega affords the earth about one 200-millionth the heat of an ordinary candle one meter distant. He even advances the dubious argument that since stars can impress their light upon a photographic plate, plants may also require the same light to carry out their nighttime activities—quite a nimble leap of illogic from the fact that film can *record* to the inference that living matter *needs*.

But Wallace didn't press this feeble, speculative argument. Instead, he emphasized that life depends upon the detailed physical structure of the universe for the same reason that Dyson cites in his two major examples: the evolution of complex, intelligent life requires a central sun that can burn steadily for untold ages, and such stable suns develop only within a delicate and narrow range of physical laws and conditions. Dyson emphasizes stellar density and diprotons; Wallace argued that appropriate suns could only exist in a universe structured like ours and only at the edge of a central cluster in such a universe.

In Wallace's universe, stars are concentrated in three regions: the outer ring (or Milky Way proper), the inner ring surrounding the central cluster, and the central cluster itself. The outer ring of the Milky Way is too dense and active a region for stable suns. Stars move so rapidly and lie so close to each other that collisions and near approaches will inevitably disrupt any planetary system before intelligent life evolves.

Wallace then claims that solar stability cannot (as we believe today) arise as a product of a star's own fuel supply (he knew little of radioactivity and nuclear fusion). Stars can burn steadily only if they are constantly supplied with new matter flowing from elsewhere. This matter moves, by gravitation, from outer regions of the universe (particularly from the ring of the Milky Way) toward the center, where our sun resides. The inner ring cannot harbor stable suns, since too much extraneous matter bombards it. The center of the solar cluster won't do, because it receives too little nurturing material. Only at the outer edge of the solar cluster, where (and surely by design) our sun resides, can a star obtain the proper balance of material to burn steadily for enough time to foster the evolution of intelligence.

Every detail of cosmic design conspires to permit life on a planet circling such a fortunately situated sun. We need the Milky Way to supply external fuel. We need the inner ring as a filter, allowing just the right amount of fuel to pass through. We need a central cluster where stars move slowly and do not interfere with each other. Could all this have happened without some directing intelligence? Eighty years after Wallace's book, our universe could not be more radically different, yet human hope continues to impose the same invalid argument upon it.

A final, important difference separates Wallace from Dyson and most modern supporters of the anthropic principle. Our contemporary advocates develop their arguments and then present their conclusion—that mind designed the universe, in part so that intelligent life might evolve within it—as a necessary and logical inference. Wallace was far too good a *historical* scientist to indulge in such fatuous certainty; he understood only too well that ordered and complex outcomes can arise from accumulated improbabilities. He therefore recognized and presented forthrightly the alternative interpretation:

> One considerable body, including probably the majority of men of science, will admit that the evidence does apparently lead to this conclusion, but will explain it as due to a fortunate coincidence. There might have been a hundred or a thousand life-bearing planets, had the course of evolution of the universe been a little different, or there might have been none at all.

This fine scientist, wearied by age and by so many lonely battles for idiosyncratic causes, but still incisively self-critical, then presented his favored interpretation, honestly recognizing its basis in a comforting view of life that could not be proved:

> The other body, and probably much the larger, would be represented by those who, holding that mind is essentially superior to matter and distinct from it, cannot believe that life, consciousness, mind, are products of matter. They hold that the marvellous complexity of forces which appear to control matter, if not actually to constitute it, are and must be mind-products.

I cannot deny that this second view, the anthropic principle, is a *possible* interpretation of the evidence, although I favor the first explanation myself. (Always be suspicious of conclusions that reinforce uncritical hope and follow comforting traditions of Western thought.) I do not object to its presentation and discussion, so long as its status as a possible interpretation, not a logical inference, receives proper identification—as Wallace did eighty years ago, and Dyson did not in our own time. I, for one, will seek my hope elsewhere. I would also be surprised, but not in the slightest displeased, if, *mirabile dictu,* Wallace and Dyson were right after all.

Postscript

Several readers informed me (as I should have remembered) that Mark Twain's famous essay, "The damned human race," was written as an explicit response to Wallace's version of the anthropic principle. Part 1 of this series, entitled "Was the world made for man?," carries as its epigraphic quote: "Alfred Russell [sic for Russel] Wallace's revival of the theory that this earth is at the centre of the stellar universe, and is the only habitable globe, has aroused great interest in the world." Twain, in his inimitable manner, then retells the history of life in five pages, assuring us that all the rich and unpatterned diversity could only represent a long pageant of preparation for that geological final second of human habitation!—so much for Wallace's assertion that the universe must have been designed with us in mind.

I was fascinated to read how many other themes of these essays lie embedded in Twain's succinct satire. For example, he explicitly cites Kelvin as his authority for the earth's *great* age—an affirmation of my argument (see essay 8) that Kelvin's work, in his own day and contrary to the common myth portraying him as an arrogant villain against empirical science, was interpreted as proof of the earth's comfortable antiquity, not as a constraint upon the immensity of time: "According to these [Kelvin's] figures, it took 99,968,000 years to prepare the world for man, impatient as the Creator doubtless was to see him and admire him. But a large enterprise like this has to be conducted warily, painstakingly, logically."

Mark Twain's ending presents a wonderful metaphor (literature and popular science contain so many) for the earth's great age relative to the length of human habitation. (I view it as a kind of literary ancestor to John McPhee's image in *Basin and Range*—that if we envision geological time as the old English yard, the distance from the King's nose to the tip of his outstretched arm, one stroke of a file applied to the nail of his middle finger would erase all of human history):

> Such is the history of it. Man has been here 32,000 years. That it took a hundred million years to prepare the world for him is proof that that is what it was done for. I suppose it is. I dunno. If the Eiffel Tower were now representing the world's age, the skin of paint on the pinnacle-knob at its summit would represent man's share of that age; and anybody would perceive that that skin was what the tower was built for. I reckon they would, I dunno.

27 | SETI and the Wisdom of Casey Stengel

SINCE THE STUDY of extraterrestrial life lacks any proven subject, opinions about the form and frequency of nonearthly beings record the hopes and fears of speculating scientists more than the constraints of evidence. Alfred Russel Wallace, for example, Darwin's partner in the discovery of natural selection and the first great evolutionist to consider exobiology in any detail, held firmly that man must be alone in the entire cosmos—for he could not bear the thought that human intelligence had not been the uniquely special gift of God, conferred upon an ideally suited planet. He wrote in 1903 that the existence of abundant and brainy extraterrestrials "would imply that man is an animal and nothing more, is of no importance in the universe, needed no great preparations for his advent, only, perhaps, a second-rate demon, and a third or fourth-rate earth" (see previous essay for a full version of this quotation and discussion of Wallace's views).

The endless debate about extraterrestrial life has focused upon the calculation of probabilities—how many stars, how many suitable planets, the chance that life will originate on appropriate earths, the probability that life will eventually generate intelligence. I must confess that I have always viewed this literature as dreary and inconclusive, too mixed up with hope and uncertainty to reach any respectable conclusion.

Recently, several astronomers and astrophysicists have

403

404 | THE FLAMINGO'S SMILE

advocated a different approach—a direct search for the
technological byproducts of intelligence by scanning the
skies systematically with radiotelescopes, probing for signals
emitted by other civilizations. This so-called SETI program
(search for extraterrestrial intelligence) has been vigorously
debated. Proponents claim that it would require but a minute fraction of the annual NASA budget and, whatever its
chances of success, would at least move the subject from
fruitless debate about probabilities to an experimental
probe by the only means now available. Opponents counter
that the scheme is a boondoggle, still costing millions and
so virtually assured of failure that it merits not a penny of
sparse public funds for science.

As an evolutionary biologist, I have no expert knowledge
in most areas motivating this debate. I am moved to comment only because opponents of SETI have featured an
argument from my field as one of their most powerful weapons. They state that all leading evolutionary biologists have
proclaimed the existence of extraterrestrial life as nearly
inconceivable. The optimism of some physical scientists
therefore resides in their failure to understand the distinctive character of evolutionary reasoning. But opponents of
SETI have misstated the biological argument, and I would
like to explain why at least one evolutionary biologist thinks
that SETI is a long shot well worth trying.

Frank J. Tipler, a mathematical physicist from Tulane
University, has been the most indefatigable critic of SETI.
In a long series of strongly worded articles for both technical and popular journals (*New Scientist, Mercury, Physics Today,
Quarterly Journal of the Royal Astronomical Society,* for example),
he gives "two basic reasons for my disbelief in the existence
of extraterrestrial intelligent beings" (all quotes from his
1982 article in bibliography, though Tipler pursues the
same themes in all his writings on SETI).

The second reason lies outside my field and I shall not
dwell on it, though it must be mentioned. Tipler argues that
"if 'they' existed, they would already be here. . . . Because
they are not here, no such beings exist." In short, Tipler
claims that any truly intelligent creatures would search or
colonize the cosmos with a device that he calls a von Neu-

mann machine—"a computer with intelligence close to the human level, capable of self-replication and capable, indeed, of constructing anything for which it has plans, using the raw materials available in the solar system it is aimed at." Intelligent life could therefore explore an entire galaxy "for the price of one von Neumann machine"—for this computer would mine asteroids and comets for material to build replicas of itself and its enclosing probe. These replicas would then scurry off to other suitable stars and replicate again. In a mere 300 million years, a whole galaxy could be saturated with the duplicated products of one von Neumann machine.

Such a machine could even fabricate the flesh and blood of extraterrestrials by mining the needed chemicals and then running the genetic program of its creator from stored memory:

> This information could in principle be stored in the memory of a von Neumann machine, which could be instructed to synthesize an egg and place the "fertilized cell" in an artificial womb. . . . In nine months there would be a human baby in the stellar system, and this could be raised to adulthood by surrogate parents, constructed by the von Neumann machine.

I don't mean to be a philistine, but I must confess that I simply don't know how to react to such arguments. I have enough trouble predicting the plans and reactions of people closest to me. I am usually baffled by the thoughts and accomplishments of humans in different cultures. I'll be damned if I can state with certainty what some extraterrestrial source of intelligence might do. Thus, Tipler's second argument follows the speculative tradition that SETI, with its experimental approach, is designed to transcend.

As his first argument, however, Tipler features a different kind of claim based on the methods and data of my field. He writes:

> First, all the great contemporary experts in the theory of evolution—Francisco Ayala, Theodosius Dob-

zhansky, Ernst Mayr, and George Simpson—are unanimous in claiming that the evolution of an intelligent species from simple one-celled organisms is so improbable that we are likely to be the only intelligent species ever to exist.

On the most mundane level, if I may play the irrelevant "expert game" for just one sentence, Tipler's statement is empirically false. I count at least four quite respectable evolutionists in the international pro-SETI petition recently released by Carl Sagan (Tom Eisner of Cornell, Dave Raup of the University of Chicago, Ed Wilson of Harvard, and with apologies for arrogance, yours truly). Evolutionary biologists, in their usual consistency with nature's primary theme, maintain a *diversity* of views on this subject.

More importantly, I think that Tipler has misunderstood what evolutionary biologists dismiss with such forcefulness by conflating two very different issues. All evolutionists who have discussed exobiology at length have clearly delineated two separate concerns—a specific claim and a general argument.

The *specific* issue considers *detailed repeatability* of any particular evolutionary sequence—in this case the evolution of creatures looking pretty much like us: bilaterally symmetrical with sense organs up front, two eyes, a nose in the middle, a mouth, and a brain. If we could start the earth's tape anew, would intelligent creatures evolve again in this form? If other worlds share our basic chemistry and conditions, would such "humanoids" evolve on them?

The *general* question asks whether attributes that we would identify as intelligence might arise in creatures of any conformation—blobs, films, spheres of pulsating energy, or diffuse and unimagined forms far beyond the limited visions of most science fiction writers.

All evolutionists have vociferously denied the specific claim, and I join them in all their vigor. Many evolutionists have also gone a step further to doubt the general argument as well, but never with such certainty—and always as a personal opinion, not as a proclamation bearing the indelible imprimatur of "evolutionary theory." I stand among those evolutionists who deny the specific claim but feel that no strong opinion can be entertained about the general argument. SETI only needs the general argument to bolster its case for support.

Gregory Bateson, the recently deceased guru of sciences that deal with complex objects and interacting systems, often emphasized that confusion of hierarchical categories may be the most common and serious fallacy of human reasoning (see his book *Mind and Nature,* for example). As a primary example of "category confusions," Bateson identified the substitution of individuals for classes (or vice versa).

Casey Stengel, one of the greatest general gurus of our time, consciously committed Bateson's fallacy of categories to avoid the heat of scrutiny in a tough moment. He was roundly criticized for blowing the Mets's first pick in the expansion draft on a *particular* catcher of quite modest ability (one Hobie Landrith by name). Casey answered by invoking the *class* of catchers in general—"You have to have a catcher, because if you don't, you're likely to have a lot of passed balls." Now Ol' Case, as usual, knew exactly what he was saying (never let the patter known as "Stengelese" fool you). He used humor to blunt criticism because he knew that we would all recognize the fallacy of reasoning and laugh at the conflation. But we commit the same error in subtler circumstances and fail to identify our confusion.

When we use "evolutionary theory" to deny categorically the possibility of extraterrestrial intelligence, we commit the classic fallacy of substituting specifics (*individual* repeatability of humanoids) for classes (the probability that evolution elsewhere might produce a creature in the *general* class of intelligent beings). I can present a good argument from "evolutionary theory" against the repetition of anything like a human body elsewhere; I cannot extend it to the general proposition that intelligence in some form might pervade the universe.

Physical scientists, following the stereotype of science as a predictable, deterministic enterprise, have often reasoned that if humans arose on earth, then we must infer (since cause leads inexorably to effect) that intelligent creatures of roughly human form would arise on any planet beginning with physical and chemical conditions similar to those that prevailed on the early earth. Perhaps this deterministic outlook is responsible for the paltry imagination of film makers and science fiction writers, with their endless creatures, all designed on a human model with two eyes, a nose, a mouth, two arms, and two legs (*Close Encounters, ET,* and even the more imaginative *Star Wars*). This tendency could be forgiven when human actors had to play the roles in our movies, but now that pieces of plastic can evoke our deepest emotions and move so subtly that ET becomes a national hero, this excuse no longer holds.

But styles of science are as diverse as their subject matter. Classical determinism and complete predictability may prevail for simple macroscopic objects subject to a few basic laws of motion (balls rolling down inclined planes in high school physics experiments), but complex historical objects do not lend themselves to such easy treatment. In the history of life, all results are products of long series of events, each so intricately dependent upon particular environments and previous histories that we cannot predict their future course with any certainty. The historical sciences try to explain unique situations—immensely complex historical accidents. Evolutionary biologists, as historical scientists, do not expect detailed repetition and cannot use the actual results of history to establish probabilities for recurrence (would a Caesar again die brutally in Rome if we could go back to *Australopithecus* in Africa and start anew?). Evolutionists view the origin of humans (or any particular butterfly, roach, or starfish) as a historical event of such complexity and improbability that we would never expect to see anything exactly like it again (or elsewhere)—hence our strong opposition to the *specific* argument about humanoids on other worlds. Consider just two of the many reasons for uniqueness of complex events in the history of life.

1. *Mass extinction as a key influence upon the history of life on earth* (see essays in section 8). Dinosaurs died some 65 million

years ago in the great worldwide Cretaceous extinction that also snuffed out about half the species of shallow water marine invertebrates. They had ruled terrestrial environments for 100 million years and would probably reign today if they had survived the debacle. Mammals arose at about the same time and spent their first 100 million years as small creatures inhabiting the nooks and crannies of a dinosaur's world. If the death of dinosaurs had not provided their great opportunity, mammals would still be small and insignificant creatures. We would not be here, and no consciously intelligent life would grace our earth. Evidence gathered since 1980 (see essay 29) indicates that the impact of an extraterrestrial body triggered this extinction. What could be more unpredictable and unexpected than comets or asteroids striking the earth literally out of the blue. Yet without such impact, our earth would lack consciously intelligent life. Many great extinctions (several larger than the Cretaceous event) have set basic patterns in the history of life, imparting an essential randomness to our evolutionary pageant.

2. *Each species as a concatenation of improbabilities.* Any animal species—human, squid, or coral—is the latest link of an evolutionary chain stretching through thousands of species back to the inception of life. If any of these species had become extinct or evolved in another direction, final results would be markedly different. Each chain of improbable events includes adaptations developed for a local environment and only fortuitously suited to support later changes. Our ancestors among fishes evolved a peculiar fin with a sturdy, central bony axis. Without a structure of this kind, landbound descendants could not have supported themselves in a nonbuoyant terrestrial environment. (Most lineages of fishes did not and could not evolve terrestrial descendants because they lacked fins of this form.) Yet these fins did not evolve in anticipation of future terrestrial needs. They developed as adaptations to a local environment in water, and were luckily suited to permit a new terrestrial direction later on. All evolutionary sequences include such a large set

of *sine quibus non,* a fortuitous series of accidents with respect to future evolutionary success. Human brains and bodies did not evolve along a direct and inevitable ladder, but by a circuitous and tortuous route carved by adaptations evolved for different reasons, and fortunately suited to later needs.

The improbabilities of history proclaim that all species are unique and unrepeatable in detail. Evolutionary theory, as a science of history, does deny the specific argument for humanoids on other worlds. All leading evolutionists, in their writings on exobiology, have said so with gusto, and I agree. Wallace began the theme in 1903:

> The ultimate development of man has, therefore roughly speaking, depended on something like a million distinct modifications, each of a special type and dependent on some precedent changes in the organic and inorganic environments, or in both. The chances against such an enormously long series of definite modifications having occurred twice over . . . are almost infinite.

Simpson has expressed the theme most eloquently in recent years, in his famous essay on "the nonprevalence of humanoids" (see bibliography):

> This essential nonrepeatability of evolution on earth obviously has a decisive bearing on the chances that it has been repeated or closely paralleled on any other planet. The assumption, so freely made by astronomers, physicists, and some biochemists, that once life gets started anywhere, humanoids will eventually and inevitably appear is plainly false. . . . Let us grant the unsubstantiated claim of millions or billions of possible planetary abodes of life; the chances of such historical duplication are still vanishingly small.

But all these evolutionists have also clearly distinguished this specific proposition about humanoids from the general argument that intelligence in some other form might arise elsewhere. On the general proposition, they have maintained a diversity of opinions—leading to the empirical conclusion that "evolutionary theory" has no clear pronouncement to make. Both Wallace and Simpson extended their argument to doubt the general claim as well, but ever so much more gently, and as a personal opinion only. Simpson, for example, wrote:

> Even in planetary histories different from ours might not some quite different and yet comparably intelligent beings . . . have evolved? Obviously these are questions that cannot be answered categorically. I can only express an opinion . . . I think it extremely unlikely that anything enough like us for real communication of thought exists anywhere in our accessible universe.

Other evolutionists, however, including two cited by Tipler as denying any possibility for SETI's success, also distinguish the specific from the general argument, but express far more optimism for the generality. Dobzhansky and Ayala, in a leading textbook (coauthored with G.L. Stebbins and J.W. Valentine), write (see bibliography):

> Granting that the possibility of obtaining a man-like creature is vanishingly small even given an astronomical number of attempts . . . there is still some small possibility that another intelligent species has arisen, one that is capable of achieving a technological civilization.

I am not convinced that the possibility is so small.

Does evolutionary theory offer any insight about the general argument? We gain some sense of probabilities for repetition of a basic theme (but not of specific details) from the phenomenon known as "convergence." Flight has evolved separately in insects, birds, pterosaurs (flying reptiles), and bats. Aerodynamic principles do not change, but morphologies differ widely (birds use feathers; bats and pterosaurs employ a membrane, but bats stretch it between several fingers, pterosaurs only from one). Marsupial "moles" and "wolves" evolved on Australia, a continent isolated from placental mammals elsewhere. Since adaptive themes are limited and animals so diverse, convergence of different evolutionary lineages to the same general solution (but not to detailed repetition) are common. Highly adaptive forms that are easy to evolve arise again and again. More complex morphologies without such adaptive necessity offer little or no prospect for repetition. Conscious intelligence has evolved only once on earth, and presents no real prospect for reemergence should we choose to use our gift for destruction. But does intelligence lie within the class of phenomena too complex and historically conditioned for repetition? I do not think that its uniqueness on earth specifies such a conclusion. Perhaps, in another form on another world, intelligence would be as easy to evolve as flight on ours.

Tipler dismisses the issue of convergence by stating that biologist Leonard Ornstein (in an article supporting Tipler, see bibliography) has refuted the most famous of all convergences—the "camera eye" of vertebrates and cephalopods (squids and their allies)—by suggesting that this structure arose in both groups from a common ancestor, and not separately in each. Even if Ornstein were right, the dismissal of a specific case does not deny the importance of convergence as a general phenomenon. But Ornstein's arguments are seriously flawed. He never mentions the strongest, "classical" argument for convergence—that the eyes, although so similar in design and operation, develop embryologically in fundamentally different ways (squid eyes form from skin precursors, while vertebrate eyes, the lens excepted, develop from the brain). Moreover, Ornstein's main argument for evolution from a common ancestor relies upon a biological principle disproved more than fifty years ago. He invokes Haeckel's discredited law that "ontogeny recapitulates phylogeny"—that an organism's embryological development repeats the sequence of ancestral adults in its evolutionary lineage. Since the eye develops so early in embryology, Ornstein argues that it may have already existed in a very remote ancestor—early enough to predate the evolutionary split of vertebrate and squid lineages. Not only has Haeckel's law been disproved (embryos do not repeat ancestral stages), but even Haeckel himself, in the heyday of his principle, rarely used time of appearance in embryology to specify moment of evolutionary origin—for he himself had identified and named a large class of exceptions to such a facile generality.

Even if we follow Tipler in arguing that von Neumann machines are the only proper way to go, he admits that we won't have the technology to build one for a century. I'm an impatient and mortal fellow. As I think it cruel to ask disadvantaged minorities to "go slow" in demands for political change—thus guaranteeing that any practical benefits will fall only upon their children's children—so too do I selfishly wish to see some exobiological results (positive or negative) in my lifetime. SETI is all we have for now. It is relatively cheap, and (in my view) entirely sensible from those perspectives that evolutionary theory can enlighten. Frankly, I think the chances of its success are a good deal lower than the probabilities envisioned by its more enthusiastic supporters among physical scientists. But we can't know until we try. Ultimately, however, I must justify the attempt at such a long shot simply by stating that a positive result would be the most cataclysmic event in our entire intellectual history. Curiosity impels, and makes us human. Might it impel others as well?

8 | Extinction and Continuity

28 | Sex, Drugs, Disasters, and the Extinction of Dinosaurs

SCIENCE, IN ITS MOST fundamental definition, is a fruitful mode of inquiry, not a list of enticing conclusions. The conclusions are the consequence, not the essence.

My greatest unhappiness with most popular presentations of science concerns their failure to separate fascinating claims from the methods that scientists use to establish the facts of nature. Journalists, and the public, thrive on controversial and stunning statements. But science is, basically, a way of knowing—in P. B. Medawar's apt words, "the art of the soluble." If the growing corps of popular science writers would focus on *how* scientists develop and defend those fascinating claims, they would make their greatest possible contribution to public understanding.

Consider three ideas, proposed in perfect seriousness to explain that greatest of all titillating puzzles—the extinction of dinosaurs. Since these three notions invoke the primally fascinating themes of our culture—sex, drugs, and violence —they surely reside in the category of fascinating claims. I want to show why two of them rank as silly speculation, while the other represents science at its grandest and most useful.

Science works with testable proposals. If, after much compilation and scrutiny of data, new information continues to affirm a hypothesis, we may accept it provisionally and gain confidence as further evidence mounts. We can

417

never be completely sure that a hypothesis is right, though we may be able to show with confidence that it is wrong. The best scientific hypotheses are also generous and expansive: they suggest extensions and implications that enlighten related, and even far distant, subjects. Simply consider how the idea of evolution has influenced virtually every intellectual field.

Useless speculation, on the other hand, is restrictive. It generates no testable hypothesis, and offers no way to obtain potentially refuting evidence. Please note that I am not speaking of truth or falsity. The speculation may well be true; still, if it provides, in principle, no material for affirmation or rejection, we can make nothing of it. It must simply stand forever as an intriguing idea. Useless speculation turns in on itself and leads nowhere; good science, containing both seeds for its potential refutation and implications for more and different testable knowledge, reaches out. But, enough preaching. Let's move on to dinosaurs, and the three proposals for their extinction.

1. Sex: Testes function only in a narrow range of temperature (those of mammals hang externally in a scrotal sac because internal body temperatures are too high for their proper function). A worldwide rise in temperature at the close of the Cretaceous period caused the testes of dinosaurs to stop functioning and led to their extinction by sterilization of males.

2. Drugs: Angiosperms (flowering plants) first evolved toward the end of the dinosaurs' reign. Many of these plants contain psychoactive agents, avoided by mammals today as a result of their bitter taste. Dinosaurs had neither means to taste the bitterness nor livers effective enough to detoxify the substances. They died of massive overdoses.

3. Disasters: A large comet or asteroid struck the earth some 65 million years ago, lofting a cloud of dust into the sky and blocking sunlight, thereby suppressing photosynthesis and so drastically lowering world temperatures that dinosaurs and hosts of other creatures became extinct.

Before analyzing these three tantalizing statements, we must establish a basic ground rule often violated in proposals for the dinosaurs' demise. *There is no separate problem of the extinction of dinosaurs.* Too often we divorce specific events from their wider contexts and systems of cause and effect. The fundamental fact of dinosaur extinction is its synchrony with the demise of so many other groups across a wide range of habitats, from terrestrial to marine.

The history of life has been punctuated by brief episodes of mass extinction. A recent analysis by University of Chicago paleontologists Jack Sepkoski and Dave Raup, based on the best and most exhaustive tabulation of data ever assembled, shows clearly that five episodes of mass dying stand well above the "background" extinctions of normal times (when we consider all mass extinctions, large and small, they seem to fall in a regular 26-million-year cycle—see essay 30). The Cretaceous debacle, occurring 65 million years ago and separating the Mesozoic and Cenozoic eras of our geological time scale, ranks prominently among the five. Nearly all the marine plankton (single-celled floating creatures) died with geological suddenness; among marine invertebrates, nearly 15 percent of all families perished, including many previously dominant groups, especially the ammonites (relatives of squids in coiled shells). On land, the dinosaurs disappeared after more than 100 million years of unchallenged domination.

In this context, speculations limited to dinosaurs alone ignore the larger phenomenon. We need a coordinated explanation for a system of events that includes the extinction of dinosaurs as one component. Thus it makes little sense, though it may fuel our desire to view mammals as inevitable inheritors of the earth, to guess that dinosaurs died because small mammals ate their eggs (a perennial favorite among untestable speculations). It seems most unlikely that some disaster peculiar to dinosaurs befell these massive beasts—and that the debacle happened to strike just when one of history's five great dyings had enveloped the earth for completely different reasons.

The testicular theory, an old favorite from the 1940s, had its root in an interesting and thoroughly respectable study of temperature tolerances in the American alligator, published in the staid *Bulletin of the American Museum of Natural History* in 1946 by three experts on living and fossil reptiles —E. H. Colbert, my own first teacher in paleontology; R. B. Cowles; and C. M. Bogert.

The first sentence of their summary reveals a purpose beyond alligators: "This report describes an attempt to infer the reactions of extinct reptiles, especially the dinosaurs, to high temperatures as based upon reactions observed in the modern alligator." They studied, by rectal thermometry, the body temperatures of alligators under changing conditions of heating and cooling. (Well, let's face it, you wouldn't want to try sticking a thermometer under a 'gator's tongue.) The predictions under test go way back to an old theory first stated by Galileo in the 1630s—the unequal scaling of surfaces and volumes. As an animal, or any object, grows (provided its shape doesn't change), surface areas must increase more slowly than volumes—since surfaces get larger as length squared, while volumes increase much more rapidly, as length cubed. Therefore, small animals have high ratios of surface to volume, while large animals cover themselves with relatively little surface.

Among cold-blooded animals lacking any physiological mechanism for keeping their temperatures constant, small creatures have a hell of a time keeping warm—because they lose so much heat through their relatively large surfaces. On the other hand, large animals, with their relatively small surfaces, may lose heat so slowly that, once warm, they may maintain effectively constant temperatures against ordinary fluctuations of climate. (In fact, the resolution of the "hot-blooded dinosaur" controversy that burned so brightly a few years back may simply be that, while large dinosaurs possessed no physiological mechanism for constant temperature, and were not therefore warm-blooded in the technical sense, their large size and relatively small surface area kept them warm.)

Colbert, Cowles, and Bogert compared the warming rates of small and large alligators. As predicted, the small fellows heated up (and cooled down) more quickly. When exposed to a warm sun, a tiny 50-gram (1.76-ounce) alligator heated up one degree Celsius every minute and a half, while a large alligator, 260 times bigger at 13,000 grams (28.7 pounds), took seven and a half minutes to gain a degree. Extrapolating up to an adult 10-ton dinosaur, they concluded that a one-degree rise in body temperature would take eighty-six hours. If large animals absorb heat so slowly (through their relatively small surfaces), they will also be unable to shed any excess heat gained when temperatures rise above a favorable level.

The authors then guessed that large dinosaurs lived at or near their optimum temperatures; Cowles suggested that a rise in global temperatures just before the Cretaceous extinction caused the dinosaurs to heat up beyond their optimal tolerance—and, being so large, they couldn't shed the unwanted heat. (In a most unusual statement within a scientific paper, Colbert and Bogert then explicitly disavowed this speculative extension of their empirical work on alligators.) Cowles conceded that this excess heat probably wasn't enough to kill or even to enervate the great beasts, but since testes often function only within a narrow range of temperature, he proposed that this global rise might have sterilized all the males, causing extinction by natural contraception.

The overdose theory has recently been supported by UCLA psychiatrist Ronald K. Siegel. Siegel has gathered, he claims, more than 2,000 records of animals who, when given access, administer various drugs to themselves—from a mere swig of alcohol to massive doses of the big H. Elephants will swill the equivalent of twenty beers at a time, but do not like alcohol in concentrations greater than 7 percent. In a silly bit of anthropocentric speculation, Siegel states that "elephants drink, perhaps, to forget . . . the anxiety produced by shrinking rangeland and the competition for food."

Since fertile imaginations can apply almost any hot idea to the extinction of dinosaurs, Siegel found a way. Flowering plants did not evolve until late in the dinosaurs' reign. These plants also produced an array of aromatic, amino-acid-based alkaloids—the major group of psychoactive agents. Most mammals are "smart" enough to avoid these potential poisons. The alkaloids simply don't taste good (they are bitter); in any case, we mammals have livers happily supplied with the capacity to detoxify them. But, Siegel speculates, perhaps dinosaurs could neither taste the bitterness nor detoxify the substances once ingested. He recently told members of the American Psychological Association: "I'm not suggesting that all dinosaurs OD'd on plant drugs, but it certainly was a factor." He also argued that death by overdose may help explain why so many dinosaur fossils are found in contorted positions. (Do not go gentle into that good night.)

Extraterrestrial catastrophes have long pedigrees in the popular literature of extinction, but the subject exploded again in 1979, after a long lull, when the father-son, physicist-geologist team of Luis and Walter Alvarez proposed that an asteroid, some 10 km in diameter, struck the earth 65 million years ago (comets, rather than asteroids, have since gained favor for reasons outlined in essay 30. Good science is self-corrective).

The force of such a collision would be immense, greater by far than the megatonnage of all the world's nuclear weapons (see essay 29). In trying to reconstruct a scenario that would explain the simultaneous dying of dinosaurs on land and so many creatures in the sea, the Alvarezes proposed that a gigantic dust cloud, generated by particles blown aloft in the impact, would so darken the earth that photosynthesis would cease and temperatures drop precipitously. (Rage, rage against the dying of the light.) The single-celled photosynthetic oceanic plankton, with life cycles measured in weeks, would perish outright, but land plants might survive through the dormancy of their seeds (land plants were not much affected by the Cretaceous extinction, and any adequate theory must account for the curious pattern of differential survival). Dinosaurs would die by starvation and freezing; small, warm-blooded mammals, with more modest requirements for food and better regulation of body temperature, would squeak through. "Let the bastards freeze in the dark," as bumper stickers of our chauvinistic neighbors in sunbelt states proclaimed several years ago during the Northeast's winter oil crisis.

All three theories, testicular malfunction, psychoactive overdosing, and asteroidal zapping, grab our attention mightily. As pure phenomenology, they rank about equally high on any hit parade of primal fascination. Yet one represents expansive science, the others restrictive and untestable speculation. The proper criterion lies in evidence and methodology; we must probe behind the superficial fascination of particular claims.

How could we possibly decide whether the hypothesis of testicular frying is right or wrong? We would have to know things that the fossil record cannot provide. What temperatures were optimal for dinosaurs? Could they avoid the absorption of excess heat by staying in the shade, or in caves? At what temperatures did their testicles cease to function? Were late Cretaceous climates ever warm enough to drive the internal temperatures of dinosaurs close to this ceiling? Testicles simply don't fossilize, and how could we infer their temperature tolerances even if they did? In short, Cowles's hypothesis is only an intriguing speculation leading nowhere. The most damning statement against it appeared right in the conclusion of Colbert, Cowles, and Bogert's paper, when they admitted: "It is difficult to advance any definite arguments against this hypothesis." My statement may seem paradoxical—isn't a hypothesis really good if you can't devise any arguments against it? Quite the contrary. It is simply untestable and unusable.

Siegel's overdosing has even less going for it. At least Cowles extrapolated his conclusion from some good data on alligators. And he didn't completely violate the primary guideline of siting dinosaur extinction in the context of a general mass dying—for rise in temperature could be the root cause of a general catastrophe, zapping dinosaurs by testicular malfunction and different groups for other reasons. But Siegel's speculation cannot touch the extinction of ammonites or oceanic plankton (diatoms make their own food with good sweet sunlight; they don't OD on the chemicals of terrestrial plants). It is simply a gratuitous, attention-grabbing guess. It cannot be tested, for how can we know what dinosaurs tasted and what their livers could do? Livers don't fossilize any better than testicles.

The hypothesis doesn't even make any sense in its own context. Angiosperms were in full flower ten million years before dinosaurs went the way of all flesh. Why did it take so long? As for the pains of a chemical death recorded in contortions of fossils, I regret to say (or rather I'm pleased to note for the dinosaurs' sake) that Siegel's knowledge of geology must be a bit deficient: muscles contract after death and geological strata rise and fall with motions of the earth's crust after burial—more than enough reason to distort a fossil's pristine appearance.

The impact story, on the other hand, has a sound basis in evidence. It can be tested, extended, refined and, if wrong, disproved. The Alvarezes did not just construct an arresting guess for public consumption. They proposed their hypothesis after laborious geochemical studies with Frank Asaro and Helen Michael had revealed a massive increase of iridium in rocks deposited right at the time of extinction. Iridium, a rare metal of the platinum group, is virtually absent from indigenous rocks of the earth's crust; most of our iridium arrives on extraterrestrial objects that strike the earth.

The Alvarez hypothesis bore immediate fruit. Based originally on evidence from two European localities, it led geochemists throughout the world to examine other sediments of the same age. They found abnormally high amounts of iridium everywhere—from continental rocks of the western United States to deep sea cores from the South Atlantic.

Cowles proposed his testicular hypothesis in the mid-1940s. Where has it gone since then? Absolutely nowhere, because scientists can do nothing with it. The hypothesis must stand as a curious appendage to a solid study of alligators. Siegel's overdose scenario will also win a few press notices and fade into oblivion. The Alvarezes' asteroid falls into a different category altogether, and much of the popular commentary has missed this essential distinction by focusing on the impact and its attendant results, and forgetting what really matters to a scientist—the iridium. If you talk just about asteroids, dust, and darkness, you tell stories no better and no more entertaining than fried testicles or terminal trips. It is the iridium—the source of testable evidence—that counts and forges the crucial distinction between speculation and science.

The proof, to twist a phrase, lies in the doing. Cowles's hypothesis has generated nothing in thirty-five years. Since its proposal in 1979, the Alvarez hypothesis has spawned hundreds of studies, a major conference, and attendant publications. Geologists are fired up. They are looking for iridium at all other extinction boundaries. Every week exposes a new wrinkle in the scientific press. Further evidence that the Cretaceous iridium represents extraterrestrial impact and not indigenous volcanism continues to accumulate. As I revise this essay in November 1984 (this paragraph will be out of date when the book is published), new data include chemical "signatures" of other isotopes indicating unearthly provenance, glass spherules of a size and sort produced by impact and not by volcanic eruptions, and high-pressure varieties of silica formed (so far as we know) only under the tremendous shock of impact.

My point is simply this: Whatever the eventual outcome (I suspect it will be positive), the Alvarez hypothesis is exciting, fruitful science because it generates tests, provides us with things to do, and expands outward. We are having fun, battling back and forth, moving toward a resolution, and extending the hypothesis beyond its original scope (see essay 30 for some truly wondrous extensions).

As just one example of the unexpected, distant cross-fertilization that good science engenders, the Alvarez hypothesis made a major contribution to a theme that has riveted public attention in the past few months—so-called nuclear winter (see next essay). In a speech delivered in April 1982, Luis Alvarez calculated the energy that a ten-kilometer asteroid would release on impact. He compared such an explosion with a full nuclear exchange and implied that all-out atomic war might unleash similar consequences.

This theme of impact leading to massive dust clouds and falling temperatures formed an important input to the decision of Carl Sagan and a group of colleagues to model the climatic consequences of nuclear holocaust. Full nuclear exchange would probably generate the same kind of dust cloud and darkening that may have wiped out the dinosaurs. Temperatures would drop precipitously and agriculture might become impossible. Avoidance of nuclear war is fundamentally an ethical and political imperative, but we must know the factual consequences to make firm judgments. I am heartened by a final link across disciplines and deep concerns—another criterion, by the way, of science at its best*: A recognition of the very phenomenon that made our evolution possible by exterminating the previously dominant dinosaurs and clearing a way for the evolution of large mammals, including us, might actually help to save us from joining those magnificent beasts in contorted poses among the strata of the earth.

---

*This quirky connection so tickles my fancy that I break my own strict rule about eliminating redundancies from these essays and end both this and the next piece with this prod to thought and action.

29 | Continuity

A GOLDEN BAND of mosaics rims the interior of Michelangelo's dome in Saint Peter's Basilica at the Vatican. It is emblazoned with that ultimate geological pun, Christ's words to Peter, taken ever since as the justification for papal supremacy and continuity. *Tu es Petrus, et super hanc petram aedificabo ecclesiam meam,* "Thou art Peter, and upon this rock I will build my church (Matthew 16:18)." In Latin, and in other languages of Christ's time, Peter's name means rock *(petra)*—so Christ appointed his first pope by name and perhaps not without a touch of humor. (It's none of my business, of course, but I have always regarded Peter—the man who denied Christ three times, and then tried to slink out of Rome until Christ reappeared and responded with gentle admonition to his *"Domine, quo vadis?"*—as a fairly weak character to assume such a weighty responsibility.) In any case, the words in golden mosaics symbolize one of the great continuities in our fickle and ephemeral history—an institution (the papacy) that can trace its lineage for two millennia.

There is no city quite like Rome, and no institution quite like the Catholic church, for appreciating continuity—that elusive property that a paleontologist like myself must deem of intrinsic and inestimable value. If subtle tuning to deep human needs and feelings represents the best formula for continuity, then the Church of Rome wins this outsider's plaudits. At the beautiful Church of Santa Maria in Tras-

427

tevere, begun in the third century, boys play soccer in the adjoining square at dusk. As the day fades, they move into the lighted portico, under wonderful mosaics of the Virgin, and continue their game admidst the tombs of early Christians. The sacred and the profane must mix.

In the Casina Pio Quatro (Pius IV's palace) on the grounds of the Vatican, I met in January 1984, the beginning of Orwell's year, with twenty scientists from eight nations to draft a statement on "nuclear winter" that the pope might use in his speeches against nuclear war. Pius IV was a sixteenth-century pope of the powerful Medici family. His house is a Roman pleasure palace, surrounded by grottoes and terraces emblazoned with statues and bas-reliefs of Roman youths in various postures of play and merriment. The ceilings are painted with swirling designs of imaginative creatures and rather frank symbols of sex and fertility. Cherubs hold aloft the six-balled Medici shield, the symbol of temporal power, with its title befitting a worldly monarch, Pius IIII Pontifex Optimus Maximus. Here and there, almost as an afterthought, a biblical scene—Christ's baptism by John, for example—fills a space amidst the Roman motifs. Again, sacred and profane, spiritual and temporal, pleasure and contemplation—all packaged into one artistic unity, a symbol of continuity that incorporates the past and recognizes human realities of the present.

I was in Rome to discuss continuity on the grandest scale. A series of studies, performed by independent groups of scientists in several nations and checked and confirmed by leaders in the various disciplines involved, seem to be converging (despite many remaining uncertainties) upon a troubling conclusion. For all prognoses about the horrors of nuclear war, we have previously missed an important theme that makes the prospect of such a holocaust even more unthinkable. We have explored the immediate consequences of blast and fallout, but we have not appreciated the longer-term effects upon climate (months to years) produced because major explosions loft clouds of dust and soot into the atmosphere. Under a range of plausible circumstances, a pall of particles might blanket the earth, bringing sub-zero temperatures to mid-latitude summers and enveloping the earth in such darkness that agriculture might fail completely. This nuclear winter raises, for the first time, the chilling prospect that a major war might not only debilitate and decimate, bringing unparalleled human suffering in its wake, but might also lead to the total and irrevocable extinction of many plant and animal species. We humans are a hardy and well-dispersed lot, but even the possibility of our own disappearance in the aftermath of nuclear winter's worst scenario cannot be totally excluded.

Why should we be so concerned about extinction? The appalling destruction of nuclear war is enough to contemplate without this added dimension. I could offer a set of "objective" reasons. Some are practical. Corn, our most important crop, will be in trouble if we lose teosinte, its wild grassy ancestor with a limited geographical and ecological distribution in Central and South America. Teosinte hybridizes with corn (see essay 24) and forms a major reserve for the genetic variability that all species require for maintenance and evolutionary flexibility. Other reasons are more frankly aesthetic. It would be an impoverished and bleak world indeed if we encountered nothing but humans and a rat or cockroach now and again. But, for this column, which I present more as a musing on continuity than a technical account of nuclear winter, I wish to emphasize a highly personal, moral argument (not subject to proof but only to simple statement, deeply felt) that flows from my own career as a paleontologist, a student of that greatest of all natural continuities, the genealogy of life on earth.

We now have evidence, in fossils of simple cells and the mats of sediment that aggregates of these cells trap and bind, that life on earth arose at least 3.5 billion years ago. It has, since then, extended upward in time, in unbroken continuity to the present. We can all, quite literally, from moss to mayfly to hippopotamus, trace our ancestry all the way back to these beginnings. The tree is an accurate metaphor for life's history; the tip of each current twig (we humans are one) flows back through branches ever wider and sturdier to the common trunk of original cells nearly 4 billion years old.

Each extinction permanently removes a bit of this patrimony; each irrevocable death of a species expunges not merely a bit of current protoplasm but a unique pathway of history maintained for 4 billion years. Each extinction is a breach of continuity on the grandest scale. Of course, from a geological perspective measured in millions of years, extinction is inevitable, even necessary for maintaining a vigorous tree of life. We may also argue, both in the abstract and for life's actual history, that an occasional catastrophic episode of mass extinction opens new evolutionary possibilities by freeing ecological space in a crowded world.

But these geological scales are not appropriate for contemplating our own life and its immediate meaning. The potentially beneficial effect of a mass extinction on life's unpredictable rebound 10 million years down the road cannot speak to the significance of our own twig on life's tree —and we do not display cosmic vanity, but merely appropriate self-interest, when we choose to nurture and defend this particular little branch.

Ours is a small twig indeed, but remember that it runs back, by myriad branches, through 4 billion years to the central trunk itself. Our origin in Africa, and our subsequent spread throughout the world, form a complex and compelling tale expressing our continuity with the entire history of life. If we extirpate this twig directly by nuclear winter, or lose so many other twigs that our own eventually withers away, then we have canceled forever the most peculiar and different, unplanned experiment ever generated among the billions of branches—the origin, via consciousness, of a twig that could discover its own history and appreciate its continuity.

Some people, who have never extricated themselves from the chain of being (see essays 17–19), and who view life's history as a tale of linear progress leading predictably to the evolution of consciousness, might be less troubled (in some abstract sense) by our potential self-removal. After all, evolution moves toward complexity and consciousness. If not us, then some other surviving branch will enter the stream and eventually give intelligence a second chance. And if not here, then elsewhere in a populated universe, for nature's laws do not vary from place to place.

As a student of life's history, and as a man who has tried hard to separate cultural prejudice and psychological hope from the story that fossils are trying to tell us, I have reached quite a different conclusion, shared, I think, by most professional colleagues: consciousness is a quirky evolutionary accident, a product of one peculiar lineage that developed most components of intelligence for other evolutionary purposes (see essay 27). If we lose its twig by human extinction, consciousness may not evolve again in any other lineage during the 5 billion years or so left to our earth before the sun explodes. Through no fault of our own, and by dint of no cosmic plan or conscious purpose, we have become, by the power of a glorious evolutionary accident called intelligence, the stewards of life's continuity on earth. We did not ask for this role, but we cannot abjure it. We may not be suited for such responsibility, but here we are. If we blow it (quite literally), we will permanently rupture a continuity of eons that dwarfs our own puny history to geological insignificance, but that we nonetheless now control. I cannot imagine anything more vulgar, more hateful, than the prospect that a tiny twig with one peculiar power might decimate a majestic and ancient tree, whose continuity stretches back to the dawn of earth's time and whose trunk and branches house so many thousand prerequisites to the twig's existence.

The argument for nuclear winter has several sources and parents. But it came to prominence toward the end of 1983 primarily through the work of a team with the appropriate acronym of TTAPS—R. P. Turco, O. B. Toon, T. P. Ackerman, J. B. Pollack, and Carl Sagan. Climatic modeling represents an unfamiliar style of science, quite different from the schoolboy stereotype of simple experiment with clear prediction and unambiguous test. We must deal, instead, with a score of variables whose values we cannot specify exactly and whose interactions are largely unknown since the experiment, thank God, has not been tried. How much dust and soot goes up; does it spread to a homogeneous layer or does it leave holes for intermittent sunlight; does it spread to the Southern Hemisphere and, if so, how intensely; where in the atmosphere do dust and soot lodge and how long will they stay before rains scavenge the particles and bring them to earth; how cold will it get; how long will the effects last? I could go on forever but will stop here. Moreover, these are only the first-order questions about unknown immediate results. What about interactions among the effects, for such "synergisms" often are, in technical parlance, nonadditive— that is, bad plus bad may not equal twice as bad, but many times worse. Radiation, for example, weakens the human immune system. It also engenders high mutation rates that might lead to the evolution of a particularly virulent agent of disease. The interaction of this new disease vector with human bodies of markedly reduced resistance might produce a pandemic far greater in effect than any prediction, based on components considered separately, could envision.

In the face of these difficulties and uncertainties, the TTAPS team proceeded by specifying the most reasonable range of values for each effect and by modeling hundreds of possible scenarios to obtain some sense of plausible scope. Major variations depend largely upon differing behaviors and amounts of dust and soot. In short, and with some simplification, direct impacts away from cities raise large amounts of fine dust high into the atmosphere; explosions over cities and forests may ignite gigantic plumes of fire that place clouds of coarser soot into lower atmospheric levels. Dust and soot block sunlight and engender nuclear winter. (I have not even mentioned scores of other profoundly negative effects, radiation and depletion of the ozone layer, for example.)

I cannot begin to handle the technical details in this short essay (the original TTAPS report and accompanying commentary by biologists, first published as two articles in *Science*, December 23, 1983, have been republished by W.W.

Norton as *The Cold and the Dark* by Paul R. Ehrlich *et al.*—
see bibliography). Carl Sagan also published a less technical, but still complete, account in the Winter 1983/84 issue
of *Foreign Affairs*). I will, however, mention just two general
conclusions. First, the threshold for nuclear winter can be
reached by many plausible scenarios involving an appropriate percentage of the world's megatonnage and a believable
number of bombs exploded over cities and military targets.
Second, and somewhat surprisingly, even a "small" nuclear
war might, under plausible circumstances, trigger nuclear
winter (for example, just 100 megatons, from our world
supply of approximately 10,000, if exploded over cities with
large subsequent fires and a maximal yield of soot, might
suffice).

I am not an astute observer of world politics, and I was
surprised (but quite pleased) that recognition of the possibility of nuclear winter has struck home with such force in
so many quarters. I have always regarded our old scenario,
restricted to the immediate consequences of blast and fallout, as so horrible that no increment of additional torment
should be needed to galvanize public opinion. But I now
realize, hopeful creatures that we are, how many people
lived with the pipe dream view, now dispersed, that if they
resided far enough from immediate blasts, and hunkered
down long enough in their shelters, they could soon emerge
into a shining world waiting to be rebuilt. I had also failed
to recognize that people in other nations, particularly of the
Southern Hemisphere, might have felt some personal
safety, also now dispersed, in the face of northern madness.
Nuclear winter also helps to clarify what seems to me the
near certainty that any "conquest" in nuclear war could
only become the ultimate Pyrrhic victory as an unforgiving
climate propagates its chilling effects upon any aggressor.

In any case, the argument of nuclear winter has, like its
dust cloud, spread throughout the world, bringing us all
perhaps a bit closer and uniting us against a common peril
—for the earth, like an organism, has its own continuity and
can disperse evenly the insults that it suffers. The Pontifical
Academy of Science, representing the world's most ecu-

menical institution, brought twenty of us from eight nations and more religions (and nonreligions) to the Vatican to draft a statement about nuclear winter and to meet with Pope John Paul II in an effort to develop this new argument as an effective weapon against the threat of nuclear war. In a short statement to us, the pope argued that we must prevail by combining our scientific deterrent (our best estimate of the factual consequences) with the moral deterrent that he and others could supply. And I thought of the wedding of spiritual and temporal, contemplation and sensuality, physical power and moral persuasion, all pictured on the sixteenth-century ceilings of our meeting place. Continuity will require this flexibility, this joining of all our forces.

We can also extend this theme of continuity, flexibility, and ecumenism to the generation of the argument about nuclear winter within science itself. Working out the details required the combined skills of physicists, meteorologists, chemists, biologists, experts on the mechanics of cratering, on the behavior of suspended particles. I am happy to report that one of the two major inspirations for the TTAPS group came directly from my own field of paleontology, so often viewed as an arcane discipline devoted to events of the distant past without immediate relevance to human life. I have written several essays on the impact theory of the Cretaceous extinction—the exciting idea, now just a few years old but gaining continually in force and evidence, that the extinction of dinosaurs and many other creatures 65 million years ago may have been triggered by the impact of comets or asteroids that struck the earth and left evidence for their bombardment in high levels of iridium, an element of great rarity in indigenous rocks of the earth's crust but more common in extraterrestrial bodies (see essays 28 and 30).

Luis Alvarez, the great Berkeley physicist and cofounder of the impact theory, has advocated from the start a scenario for extinction based upon a massive dust cloud thrown aloft by the cosmic crash, with subsequent suppression of photosynthesis and plunging temperatures. He also explicitly recognized the parallels between cometary zap and nuclear war (in fact the megatonnage of such an impact greatly exceeds the power of our entire nuclear arsenal). Sagan and his colleagues read the message and applied it directly. Good science also displays its continuity across apparently unrelated disciplines.

The impact potentiated our own evolution; without such a blast, I doubt that we would be here to contemplate nuclear winter. Mammals evolved at about the same time as dinosaurs and lived their first 100 million years as small creatures at the periphery of a world dominated by giant reptiles. If dinosaurs had not become extinct in the Cretaceous bombardment, they presumably would still dominate the earth (as they had for 100 million years, so why not for 65 million more), mammals would have remained as small creatures of ratlike size, and intelligence would not have evolved to create the glories of intellect and the horrors of nuclear holocaust. Is it not a hopeful thought that by recognizing the cause for a key event that inspired our own evolution, we might also contribute, through its direct input in formulating the argument of nuclear winter, to our survival against the greatest threat ever produced by the tree of life against its own fragile continuity.

————

Postscript

The official Vatican's statement, drafted at our meeting, has now been published. Its full text appears below.

NUCLEAR WINTER: A WARNING

Nuclear war would include among its immediate consequences the death of a large proportion of the populations in combatant nations. Such a war would represent a catastrophe unprecedented in human history. Subsequent radioactive fallout, weakening of the human immune system, disease, and the collapse of medical and other civil services would threaten large numbers of survivors.

We must now issue an additional warning: newly-recognized effects of nuclear war on the global climate indicate that longer-term consequences might be as dire as the prompt effects, if not worse.

In a nuclear war, weapons exploded near the ground would inject large quantities of dust into the atmosphere, and those exploded over cities and forests would suddenly generate enormous amounts of sooty smoke from the resulting fires. The clouds of fine particles would soon spread throughout the Northern Hemisphere, absorbing and scattering sunlight and thus darkening and cooling the earth's surface. Continental temperatures could fall rapidly—well below freezing for months, even in summertime—creating a "nuclear winter." This would happen even with wide variations in the nature and extent of nuclear war.

We have only recently become aware of how severe the cold and the dark might be, especially as a consequence of intense and numerous fires ignited by nuclear explosions, and from attendant changes in atmospheric circulation. This would produce a profound additional assault upon surviving plants, animals and humans. Agriculture, at least in the Northern Hemisphere, could be severely damaged for a year or more, causing widespread famine.

Calculations show that the dust and smoke may well spread to the tropics and to much of the Southern Hemisphere. Thus non-combatant nations, including those far from the conflict, could be severely afflicted. Such nations as India, Brazil, Nigeria, and Indonesia could be struck by unparalleled disaster, without a single bomb exploding on their territories.

Moreover, nuclear winter might be triggered by a relatively small nuclear war, involving only a minor fraction of the present global strategic arsenals, provided that cities are targeted and burned. Even if a "limited" nuclear war were initiated in a manner intended to minimize such effects, it would likely escalate to the massive use of nuclear weapons, as the Pontifical Academy of Sciences stressed in its earlier "Declaration on Prevention of Nuclear War" (1982).

The general results seem to be valid over a wide range of plausible conditions, and over wide variations in the character and extent of a nuclear war. However, there are still uncertainties in the present evaluations, and there are effects which have not yet been studied. Therefore, additional scientific work and continuing critical scrutiny of methods and data are clearly required. Unanticipated further dangers from nuclear war cannot be excluded.

Nuclear winter implies a vast increase in human suffering, including nations not directly involved in the war. A large proportion of humans who survive the immediate consequences of nuclear war would most likely die from freezing, starvation, disease, and, in addition, the effects of radiation. The extinction of many plant and animal species can be expected, and, in extreme cases, the extinction of most non-oceanic species might occur. Nuclear war could thus carry in its wake a destruction of life unparalleled at any time during the tenure of humans on Earth, and might therefore imperil the future of humanity.

Carlos Chagas, Brazil, *Chairman*

Vladimir Alexandrov, USSR

Edoardo Amaldi, Italy

Dan Beninson, Argentina

Paul J. Crutzen, FRG

Lars Ernster, Sweden

Giorgio Fiocco, Italy

Stephen J. Gould, USA

José Goldemberg, Brazil

S.N. Isaev, USSR

Raymond Latarjet, France

Louis Leprince-Ringuet, France

Carl Sagan, USA

Carlo Schaerf, Italy

Eugene M. Shoemaker, USA

Charles Townes, USA

Eugene P. Velikhov, USSR

Victor Weisskopf, USA

30 | The Cosmic Dance of Siva

VULCAN, THE ROMAN GOD OF FIRE, bestowed his name upon a planet for a few years during the nineteenth century. Appropriately situated in the hottest spot of our immediate heavens, between Mercury and the sun, this putative planet emerged because Newtonian science knew no other way to produce (by gravitational pull) the slight irregularity that had been measured in Mercury's orbit. Since Vulcan had to exist, and since theory can exert such a remarkable effect upon observation, several sightings were actually reported. We now understand that gravitation is Einsteinian, not perfectly Newtonian, and equations of relativity adequately explain the perturbations of Mercury without an additional disturbing body. Deprived of its theoretical necessity, Vulcan quietly disappeared.

No scientific activity teeters more precariously on the precipice between bravery and foolishness than descriptions of unobserved objects justified only by their necessity in theory. The audacious can even take a firmer step toward perdition or renown by conferring a formal name upon their hypothetical entity. What can a friendly bystander say about such a strategy? We can formulate no general rules for success; as Nick the Greek might say, "ya win some, ya lose some." The proponents of Vulcan lost big, but others have triumphed in the same game.

Ernst Haeckel, Germany's leading evolutionist in Darwin's day, drew a hypothetical lineage of human evolution thirty years before Eugene Dubois discovered the first transitional fossils. On this tree, *Homo sapiens* reached back to a less worthy predecessor named *Homo stupidus*—a hypothetical cretin itself descended from the true missing link joining apes and humans. Haeckel had no fossils, but he did have a name. He called this putative ancestor *Pithecanthropus alalus,* or the ape-man who could not speak. But Haeckel won, where the Vulcanophiles met defeat. So accurate were Haeckel's major predictions—particularly his claim that our immediate ancestor would walk fully erect but possess a brain far smaller than ours—that Dubois willingly accepted his name, christening the first human fossils *Pithecanthropus erectus* (the specimens from Java now called *Homo erectus*).

In April 1984, inspired by a new theory for the cause of mass extinctions, several scientists christened another unobserved member of our solar system. The sun, they proposed, has a previously unrecognized companion star, revolving in an eccentric orbit and now at a maximal distance of more than two light-years (hence, at its small mass and low luminosity, so barely discernible, even with the most powerful telescopes, that we would easily and forever miss it unless searching directly). They also—why not go whole hog while you're at it—proposed a name for the sun's hypothetical companion. They have called it Nemesis (I shall explain why in a moment) to honor the Greeks' personification of righteous anger in the form of a goddess. "We worry," they wrote, "that if the companion is not found, this paper will be our nemesis" (Marc Davis, Piet Hut, and Richard A. Muller, see bibliography. Daniel P. Whitmire and Albert A. Jackson IV independently postulated the existence of Nemesis in the same issue of *Nature*).

The prediction of Nemesis culminates a long series of disparate discoveries and conjectures, spanning more than a century but gathering considerable steam in the past few months. I have discussed each item, often many times, during a decade of essays. Their current conjunction and synthesis either marks the most exciting event in my profession of paleontology during my lifetime or just another mistake made by those fallible mortals known as scientists. (I now have ten pounds riding on excitement with a skeptical English colleague.) With my lead time of three months* and the spate of preemptive articles in newspapers and magazines produced more quickly, I would be performing no service in presenting a straight exposition of the theory itself. I wish, instead, to explain why this new theory of mass extinction might be so vitally important in altering our basic conception of the causes of pattern in life's history. I also want to end with a little gloss upon the theory itself—a plea to potential spotters that they name our companion Siva, not Nemesis, both to express the ecumenical spirit of science at its best and to recognize an almost devastating match between the proposed role of a solar companion in mass extinction and the attributes of this Eastern god of destruction. But first, let me list the primary events now coalescing into a new view of mass extinction.

1. Geologists have known for nearly two centuries that extensive extinctions, affecting life in a wide range of environments, have occurred sporadically and rapidly many times during the past 600 million years. Our geological time scale depends upon these mass extinctions since they set the boundaries of major divisions. My standard response to generations of student groans (at the imposed necessity of memorizing all those funny names from Cambrian to Pleistocene) reminds my charges that they are not learning capricious words for the arbitrary division of continuous time, but rather the dates of major events in the history of life.

2. Theories of mass extinction would fill a book thick enough to prop any junior to adult height at the dinner table. But an impasse broke some five years ago, when high levels of iridium in rocks at the Cretaceous-Tertiary boundary (dinosaur doomsday) provided the first solid evidence for coincidence between extraterrestrial impact and times of extinction (see essay 25 in *Hen's Teeth and*

---

*Now up to one-and-a-half years between composition and book—an impossibly long time for an exciting area in science.

Horse's Toes). Iridium is a heavy, unreactive element, and the earth's original supply presumably sank into its interior when our planet melted and differentiated some 4 billion years ago. Iridium in surface rocks arrives largely from extraterrestrial sources—asteroids, meteorites, and comets. Unless, of course, the earth's original iridium can rise from the interior in volcanic eruptions—the only serious challenge proposed against the impact theory.

3. Luis Alvarez, Walter Alvarez, Frank Asaro, and Helen Michel proposed that a large asteroid, some ten kilometers in diameter, struck the earth and dumped the iridium some 65 million years ago. They based their suggestion on enhanced iridium in three sites, all for one extinction. Paleontological reactions ranged initially from skepticism to derision (I take considerable pride, in a career liberally studded with error, in my iconoclastic original enthusiasm). Since then, the tenuous base of initial evidence has been greatly strengthened. Enhanced iridium has been found throughout the world in more than fifty localities right at the Cretaceous-Tertiary boundary, from terrestrial sediments to deep-sea cores. Iridium has also been discovered, with varying degrees of confidence, in rocks marking four or five other episodes of mass extinction.

4. David Raup and Jack Sepkoski, working from extensive compilations of the life and death times for fossil families, found a 26-million-year periodicity in extinctions during the past 225 million years (see essay 15). (This cyclicity had not been noted before because the smallest of these extinctions could not be separated from ordinary background levels before Sepkoski compiled his more extensive and refined data.)

5. Walter Alvarez and Richard A. Muller found a periodicity, similar in timing and spacing (28.4 million years), to the Raup-Sepkoski extinction peaks, for well-dated impact craters on earth with diameters in excess of ten kilometers. Since such craters are rare (fewer than twenty), conclusions must be tentative, but the coincidence of two data sets—neither presumed in the past

either to show cyclicity or (for that matter) to have anything to do with each other—is (to say the least) sugges-
tive.

6. So far, so solid. The rest is productive speculation about
mechanisms: Cyclicity undermined the Alvarez asteroid
(good science is self-corrective). Asteroidal impacts, as
we understand them, can only occur at random when a
so-called Apollo object (an asteroid with an orbit eccentric enough to traverse our part of the sky during its
wanderings) strikes the earth. What extraterrestrial object could bring in iridium but also hit the earth with
consistent rhythm? Thought shifted to comets.

7. Second-level speculation: Billions of comets circle the
sun in an envelope called the Oort cloud and located well
beyond the orbit of Pluto. Gravitational disturbance of
this cloud might alter cometary orbits and send large
numbers hurtling into the space of the inner planets.
Some would then strike the earth.

8. Third-level speculation: What could so perturb the Oort
cloud at a 26-million-year periodicity? Various suggestions have emerged. Oscillations of the solar system with
respect to the plane of our galaxy (bringing the Oort
cloud in and out of contact with interstellar clouds of
dust and gas) have been proposed, but the timing and
length of these excursions—a cycle of some 33 million
years—fit the extinction and cratering data poorly. A
solar companion, on an orbit so eccentric that it perturbs
the Oort cloud only on its closest approach, seems to
work in principle. Such a notion sounds, I freely confess,
like science fiction of the lowest order, but the idea must
be taken most seriously, for it obeys the cardinal criterion of fruitful science. It is plausible in theory and testable in practice (see essay 28). We can scan the skies and
hope to know—a gamble well worth taking (even for low
probability), given the immense intellectual reward of
potential success. Piet Hut told me that we should have
a 50 percent chance of finding the companion within
three years, if it exists. And, oh yes, don't worry. Our

companion is now at its maximal distance; the Oort cloud won't be jolted for another 13 million years or so.

Cometary showers and shrouds of dust must titillate anyone's fancy, but their fascination for paleontologists lies not with the wham-bam of the western movie scenario, but in a profound implication that we must face squarely and that may fundamentally alter our favorite principle for explaining life's history. We may identify two extreme (and contrasting) positions as guides for interpreting life's pattern in time. (All astute paleontologists recognize that the truth lies somewhere in between, but I wish to argue that the first has been favored as a kind of controlling metaphor, while new views on mass extinction suggest a far greater role for the second.)

The first holds that competition among species drives the history of life forward and specifies its steady changes. Even if environments were perfectly constant, evolution would continue as organisms struggle (literally or figuratively) with others in the race for life. You don't necessarily get anywhere (measured by triumph over others) because everyone else is struggling too, but the net result is a kind of upward relay preserving balances among competitors as all struggle for temporary advantages. Paleontologist Leigh Van Valen has codified this model for life's history as the "Red Queen" hypothesis in honor of Alice's compatriot (in *Through the Looking Glass*), who had to keep running all the time just to stay in the same place.

The Red Queen has been our dominant model for life's history. It is Darwin's own controlling metaphor of the wedge recast for the fullness of time:

Nature may be compared to a surface covered with ten-thousand sharp wedges . . . representing different species, all packed closely together and driven in by incessant blows, . . . sometimes a wedge of one form and sometimes another being struck; the one driven deeply in forcing out others; with the jar and shock often transmitted very far to other wedges in many lines of direction.

Nature, in other words, is always full (or near equilibrium, in technical parlance). One form can gain a space only by pushing another out ("wedging," in Darwin's words). The metaphor of the wedge underlies and supports our conventional view of life's order: Creatures strive to improve themselves; life moves steadily upward although no one gets permanently ahead; order rules as the predictable struggle of individuals translates to patterns of increasing complexity and diversity. Marx was not far wrong when he remarked of Darwin's system that it resembled Hobbes's *bellum omnium contra omnes* (the war of all against all) imposed upon nature.

The second, or minority, view holds that no internal dynamic drives life forward. If environments did not change, evolution might well grind to a virtual halt. At a high level of paleontological resolution (if not among the bugs and birds in my garden), species pass their lives in general independence, as Longfellow's "Ships that pass in the night.... Only a signal shown and a distant voice in the darkness." Their primary "struggles" are with changing climates, geologies, and geographies, not with each other. (Competition then becomes a sporadic and local interaction, smoothing and shaping the edges of life's order, but not acting as its driving force.)

In this view, external triggers of changing environment must drive the history of life forward. But they drive it in unconventional directions: Where can we find the upward advance that we seek so assiduously (to put ourselves on top of a struggling mass) if life only tracks a capriciously changing environment? Where can we locate predictable order at all if the primary environmental triggers are periodic cometary showers?

To cite a specific example contrasting the two views and their differing implications, I restudied (with C. Brad Calloway, see bibliography) the standard textbook case of wedging on a grand scale: the interaction of clams and brachiopods through time. These major groups of marine inverte-brates look superficially similar: both cover their body with two shells and most species either attach to the sea bottom or, with limited mobility, burrow into sediments. But clams have a more complex anatomy and are conventionally ranked higher in the old Procrustean classifications that forced the bush of life into linear order. Clams also dominate many marine faunas today, while brachiopods are relatively inconspicuous; our early fossil record, however, is replete with brachiopods and depauperate in clams. Thus, we have all the ingredients for a classic tale of gradual competitive replacement by wedging—superior clams, step by step, force brachiopods out of their limited, mutual environment. Calloway and I gathered a compendium of statements, span-ning more than a century, and all citing clams and brachio-pods as the classic case of progress in life's history by com-petitive exclusion.

But we found that the numbers don't support this facile tale. Clams and brachiopods do not show the fine-scale negative interaction that wedging requires. In fact, they vary in sympathy throughout geological time: periods with more than an average number of clams are enriched in brachiopods as well; stages deprived of brachiopods are also weak in clams. Moreover, each group seems to follow its own distinctive course in normal times, oblivious to the other's fate and history: clams increase slowly within each chunk of normal time; brachiopods hold their own.

The old story represents a false inference from one basic fact: brachiopods do dominate early faunas, while clams are so abundant today that Ho Jo can feed a nation on their breaded feet. But we found that the supposed "replace-ment" of brachiopods by clams does not occur by gradual competitive wedging, but simply records different reactions to that greatest of all mass dyings—the Permian extinction (when more than 90 percent of species probably perished). Brachiopods really took it on the (metaphorical) chin; clams scarcely noticed the debacle. Thus, clams got "ahead" of brachs in this one geological moment and never relin-quished their new incumbency. The fossil pattern records independent reactions to a single mass extinction, not gradual wedging and triumph of superior anatomies. Clams and brachiopods act like ships passing in the night, but faring differently in the great tempest.

In short, if mass extinctions are so frequent, so profound in their effects, and caused fundamentally by an extraterrestrial agency so catastrophic in impact and so utterly beyond the power of organisms to anticipate, then life's history either has an irreducible randomness or operates by new and undiscovered rules for perturbations, not (as we always thought) by laws that regulate predictable competition during normal times.

All this ferment may be disturbing to our hopes and our desires to find a sop or solace in nature, but it presents paleontology with the richest possible field for thought and action. For we students of life's history are guardians of the data that can resolve these fundamental issues. The cyclical theory of catastrophic extinction leaves paleontologists in the driver's seat with a decade of exciting work before us. Scientists rarely have the privilege of addressing such fundamental questions in a new and fruitful manner.

I cannot, in this context, present a technical program for paleontological work, but consider just three issues demanding attention and amenable to resolution from the fossil record:

1. How much of the 26 million years between catastrophes does life need to recover its former richness (in numbers of species and ecological complexity of communities)? If most time passes in periods of recovery, then competitive models must fail (since they require a full world for the wedge's metaphor) and external triggers must drive life's history.

2. Are patterns of who dies and who survives a catastrophe consistent with purely random removals from the field of life? If randomness fails, do the regularities of mass extinction record rules different from those governing the order of normal times between catastrophes? Under either a random or "different rules" model, the Darwinian

hope of smooth extrapolation from small-scale events (which can be studied directly) to the great geological panorama fails, and we must recognize the distinctive character that mass extinction imparts to life's history.

3. Why are the cyclical extinctions so different in strength (one wiping out more than 90 percent of species, others protruding so little above background that we needed Sepkoski's refined data to recognize them at all)? Some cometary enthusiasts, in the wave of overattribution that accompanies most new ideas, are trying to explain everything by impact. If perturbations of the Oort cloud send billions of comets hurtling toward the planets, only a handful will strike the earth—sometimes more, sometimes fewer. Big extinctions mean more comets; little extinctions, fewer. But it cannot be so mechanically simple. We have compiled a century of data on correlations of terrestrial events with mass extinctions (many, for example, are accompanied by falling sea levels); we also know that several extinctions were preceded by long, gradual, and simultaneous declines in many groups. We used to think that these terrestrial correlates would explain the extinctions. I suspect that we need a reversed perspective, but one that will still cherish the terrestrial data. Terrestrial correlates are probably not the causes but the primary regulators of severity. When comets hit a biosphere weakened for other reasons, unusually large extinctions ensue. The greatest of all extinctions occurred on an earth with all continents coalesced into a single Pangaea. I used to think that Pangaea was the primary cause (see essay 16 in *Ever Since Darwin*); I now think that it was the stage for maximal severity.

To end these universal bangs with a personal whimper, may I make my little suggestion to astronomical colleagues pursuing the good search. If Thalia, the goddess of good cheer, smiles upon you and you find the sun's companion star, please do not name it (as you plan) for her colleague Nemesis. Nemesis is the personification of righteous anger. She attacks the vain or the powerful, and she works for definite cause (punishing Narcissus, for example, with his burden of unquenchable self-esteem). She represents everything that our new view of mass extinction is struggling to replace—predictable, deterministic causes afflicting those who deserve it. She would also place one more Western figure into a universal sky. May not one member of our solar system honor the traditions of another culture?

Mass extinctions are not unswervingly destructive in the history of life. They represent a source of creation as well, especially if the second view of external triggering has validity, and the Red Queen of internal competition does not drive life inexorably forward. Mass extinction may be the primary and indispensable seed of major changes and shifts in life's history. Destruction and creation are locked in a dialectic of interaction. Moreover, mass extinction is probably blind to the exquisite adaptations evolved for previous environments of normal times. It strikes at random or by rules that transcend the plans and purposes of any victim. May we not name the sun's potential companion for a figure who embodies these central features of creativity in destruction and "neutrality" toward the evolutionary struggles of creatures in preceding normal times?

Siva, the Hindu god of destruction, forms an indissoluble triad with Brahma, the creator, and Vishnu, the preserver. All are enmeshed in one—a trinity of a different order—because all activity reflects their interaction. A. Parthasarathy writes in his *Symbolism of Hindu Gods and Rituals*: "All three powers are manifest at all times. They are inseparable. Creation and destruction are like two sides of a coin. . . . Morning dies to give birth to noon. Noon dies when night is born. In this chain of birth and death the day is maintained"—as the balances of life's history arise from creative recoveries following massive destructions.

Siva is often, and most beautifully, presented in the form of Nataraja, the cosmic dance. He holds in one hand the flame of destruction, in another (he has four in all) the *damaru,* a drum that regulates the rhythm of the dance and symbolizes creation. He moves within a ring of fire—the

The Hindu god Siva in the form of Nataraja. He holds the flame
of destruction in one hand, and a drum to regulate the rhythm of
the dance (and symbolize creation) in another. He moves in a ring
of fire—maintained by the interaction of creation and destruction.
THE ASIA SOCIETY, NEW YORK. MR. AND MRS. JOHN D. ROCKEFELLER
3RD COLLECTION. PHOTO BY OTTO E. NELSON.

cosmic cycle—maintained by an interaction of destruction
and creation, beating out a rhythm as regular as any clockwork of cometary collisions. "In this perpetual process of
creation and destruction," Parthasarathy writes, "the universe is maintained." Unlike Nemesis, Siva does not attack

specific targets for cause or for punishment. Instead, his placid face records the absolute tranquillity and serenity of a neutral process, directed toward no one but responsible for maintaining the order of our world.

Most hot ideas turn out to be wrong. I can only hope that I will not be remembered as the man who campaigned with a name for the nonexistent (surely worse than a moon for the misbegotten). Some chances are certainly worth taking. If Thalia smiles and Siva exists, think what it all will mean for my beloved science of paleontology. We have labored so long under the onus of boredom and dullness. We are the guardians of life's history, but we are often depicted as mindless philatelists of stone; specialists in tiny corners of space, time, and taxonomy; purveyors of such arcane names as *Pharkidonotus percarinatus* in extended orgies of irrelevant detail. The editors of Britain's leading scientific journal wrote of us in 1969: "Scientists in general might be excused for assuming that most geologists are paleontologists and most paleontologists have staked out a square mile as their life's work."

Times have been changing for more than a decade, but Siva would crown our transformation. What an apotheosis for a previously "dull" science—to be the source and impetus, by discovering the 26-million-year cycle, for the greatest revision of cosmology (at least for our little corner of the heavens) since Galileo.

Bibliography

Agassiz, L. 1862. *Contributions to the natural history of the United States.* Vol. 4. Boston.

Altick, R.D. 1978. *The shows of London.* Cambridge, MA: Harvard University Press.

Alvarez, L.W. 1982. Experimental evidence that an asteroid impact led to the extinction of many species 65 million years ago. *Proceedings of the National Academy of Sciences* 80: 627–42.

Alvarez, L.W.; W. Alvarez; F. Asaro; and H.V. Michel. 1980. Extraterrestrial cause for the Cretaceous-Tertiary extinction. *Science* 208: 1095–1108.

Alvarez, W., and R.A. Muller. 1984. Evidence from crater ages for periodic impacts on the earth. *Nature* 308: 718–20.

Anonymous. 1969. What will happen to geology. *Nature* 221: 903.

Barnes, C.W. 1980. *Earth, time and life.* New York: John Wiley.

Bateson, G. 1979. *Mind and nature.* New York: E.P. Dutton.

Beadle, G.W. 1980. The ancestry of corn. *Scientific American,* January, pp. 112–19.

Bigelow, R.P. 1900. The anatomy and development of *Cassiopea xamachana. Memoirs Boston Society of Natural History* 5: 191–236.

Briggs, D.E.G.; E.N.K. Clarkson; and R.J. Aldridge. 1983. The conodont animal. *Lethaia* 16: 1–4.

451

Buckland, W. 1823. *Reliquiae diluvianae; or, observations on the organic remains contained in caves, fissures, and diluvial gravel, and on other geological phenomena attesting the action of a universal deluge.* London: John Murray.

Buckland, W. 1836 (1841 edition). *Geology and mineralogy considered with reference to natural theology.* Philadelphia: Lea and Blanchard.

Buffon, G.L. 1828. *Oeuvres complètes de Buffon.* Edited by M.A. Richard. Vol. 28. Paris: Baudouin.

Burchfield, J.D. 1975. *Lord Kelvin and the age of the earth.* New York: Science History Publications.

Buskirk, R.E.; C. Frohlich; and K.G. Ross. 1984. The natural selection of sexual cannibalism. *American Naturalist* 123: 617–25.

Colbert, E.H.; R.B. Cowles; and C.M. Bogert. 1946. Temperature tolerances in the American alligator and their bearing on the habits, evolution, and extinction of the dinosaurs. *Bulletin of the American Museum of Natural History* 86: 327–74.

Coon, C. 1962. *The origin of races.* New York: A.A. Knopf.

Cuvier, G. 1817. Extrait d'observations faites sur le cadavre d'une femme connue à Paris et à Londres sous le nom de Vénus hottentotte. *Mémoires du Muséum d'Histoire Naturelle* 3: 259–74.

Darwin, C. 1859. *On the origin of species.* London: John Murray.

Darwin, C. 1871. *The descent of man and selection in relation to sex.* London: John Murray.

Davis, M.; P. Hut; and R.A. Muller. 1984. Extinction of species by periodic comet showers. *Nature* 308: 715–17.

Dobzhansky, T.; F.J. Ayala; G.L. Stebbins; and J.W. Valentine. 1977. *Evolution.* San Francisco: W.H. Freeman.

Dyson, F. 1979. *Disturbing the universe.* New York: Harper and Row.

Ehrlich, P.R.; C. Sagan; D. Kennedy; and W.O. Roberts. 1984. *The cold and the dark. The world after nuclear war.* New York: W.W. Norton.

Eiseley, L. 1958. *Darwin's century.* New York: Doubleday.

Garrett, P., and S.J. Gould. 1984. Geology of New Providence Island, Bahamas. *Geological Society of America Bulletin* 95: 209–20.

Goldschmidt, R. 1940 (reprinted 1982 with introduction by S.J. Gould). *The material basis of evolution*. New Haven: Yale University Press.

Gosse, P.H. 1857. *Omphalos: An attempt to untie the geological knot*. London: John Van Voorst.

Gould, S.J. 1977. *Ever since Darwin*. New York: W.W. Norton.

Gould, S.J. 1977. *Ontogeny and phylogeny*. Cambridge, MA: Belknap Press of Harvard University Press.

Gould, S.J. 1980. *The panda's thumb*. New York: W.W. Norton.

Gould, S.J. 1981. *The mismeasure of man*. New York: W.W. Norton.

Gould, S.J. 1983. *Hen's teeth and horse's toes*. New York: W.W. Norton.

Gould, S.J., and C.B. Calloway. 1980. Clams and brachiopods—ships that pass in the night. *Paleobiology* 6: 383–96.

Gould, S.J., and D.S. Woodruff. 1978. Natural history of *Cerion* VIII: Little Bahama Bank—a revision based on genetics, morphometrics, and geographic distribution. *Bulletin of the Museum of Comparative Zoology* 148: 371–415.

Grew, N. 1681. *Musaeum regalis societatis, or a catalogue and description of the natural and artificial rarities belonging to the Royal Society and preserved at Gresham Colledge, whereunto is subjoyned the comparative anatomy of stomachs and guts*. London: Thomas Malthus.

Guenther, M. 1980. The changing Western image of the Bushmen. *Paideuma* 26: 123–40.

Haeckel, E.H. 1869. *Über Arbeitstheilung in Natur und Menschenleben (On the division of labor in nature and human life)*. Berlin.

Haeckel, E.H. 1888. Report on the Siphonophorae collected by *HMS Challenger* during the years 1873–1876. Voyage of *HMS Challenger*, Zoology, Vol. 28.

Hearnshaw, L.S. 1979. *Cyril Burt, psychologist*. London: Hodder and Stoughton.

Hoagland, K.E. 1978. Protandry and the evolution of environmentally-mediated sex change: A study of the Mollusca. *Malacologia* 17: 365–91.

Howard, L.O. 1886. The excessive voracity of the female Mantis. *Science* 8: 326.

Huxley, T.H. 1849. *The oceanic Hydrozoa observed during the voyage of* HMS "Rattlesnake" *in the years 1846–1850.* London: The Ray Society.

Huxley, T.H. 1863. *Evidence as to man's place in nature.* London: Williams and Norgate.

Iltis, H.H. 1983. From teosinte to maize: The catastrophic sexual transmutation. *Science* 222: 886–94.

Jenkin, P.M. 1957. The filter feeding and food of flamingoes (Phoenicopteri). *Philosophical Transactions of the Royal Society of London,* Series B. 240: 401–93.

Jensen, A.R. 1969. How much can we boost IQ and scholastic achievement. *Harvard Educational Review* 33: 1–123.

Just, E.E. 1912. The relation of the first cleavage plane to the entrance point of the sperm. *Biological Bulletin* 22: 239–52.

Just, E.E. 1933. Cortical cytoplasm and evolution. *American Naturalist* 67: 20–29.

Just, E.E. 1939. *The biology of the cell surface.* Philadelphia: P. Blakiston's Son.

Just, E.E. 1940. Unsolved problems of general biology. *Physiological Zoology* 13: 123–42.

Kamin, L.J. 1974. *The science and politics of IQ.* Potomac, MD: Lawrence Erlbaum Associates.

Keeton, W.T. 1980. *Biological science.* New York: W.W. Norton.

Kinsey, A.C. 1930. *The gall wasp genus* Cynips: *A study in the origin of species.* Indiana University Studies, Vol. 16, 577 pp.

Kinsey, A.C. 1936. *The origin of higher categories in* Cynips. Indiana University Publications, Science Series No. 4, 334 pp.

Kinsey, A.C.; W.B. Pomeroy; and C.E. Martin. 1948. *Sexual behavior in the human male.* Philadelphia: W.B. Saunders.

Kinsey, A.C.; W.B. Pomeroy; C.E. Martin; and P.H. Gebhard. 1953. *Sexual behavior in the human female.* Philadelphia: W.B. Saunders.

BIBLIOGRAPHY | 455

Lack, D. 1947. *Darwin's finches: An essay on the general biological theory of evolution.* Cambridge, England: Cambridge University Press.

Lamarck, J.B. 1809 (reprinted 1984). *Zoological philosophy.* Chicago: University of Chicago Press. (I used the French original, so wordings of quotes will differ.)

Lewontin, R.C. 1982. *Human diversity.* New York: Scientific American Library.

Linnaeus, C. 1758 (facsimile reprint 1956). *Systema naturae. Regnum animale.* London: British Museum (Natural History).

Lovejoy, A. 1936. *The great chain of being.* Cambridge, MA: Harvard University Press.

Lyell, C. 1830–1833. *Principles of geology.* London: John Murray.

Manning, K.R. 1983. *Black Apollo of science: The life of Ernest Everett Just.* New York: Oxford University Press.

Maupertuis, P. (published anonymously). 1745. *Vénus physique.* Publisher unspecified, 194 pp.

Mayer, A.G. 1910. *Medusae of the world.* Vol. 3. Publications of the Carnegie Institute of Washington, No. 109.

Mayr, E. 1982. *The growth of biological thought.* Cambridge, MA: Harvard University Press.

Montagu, A. 1943. *Edward Tyson, M.D., F.R.S. 1650–1708, and the rise of human and comparative anatomy in England.* Memoirs of the American Philosophical Society, Vol. 20, 488 pp.

Montagu, A. 1945. Intelligence of northern Negroes and southern whites in the First World War. *American Journal of Psychology* 58: 161–88.

Morton, S.G. 1839. *Crania americana.* Philadelphia: John Pennington.

Oken, L. 1847. *Elements of physiophilosophy* (translation by A. Tulk of Oken's *Lehrbuch der Naturphilosophie*). London: Ray Society.

Ornstein, L. 1982. A biologist looks at the numbers. *Physics Today,* March, pp. 27–31.

Parthasarathy, A. 1983. *The symbolism of Hindu gods and rituals.* Bombay: Shailesh Printers.

Perkins, H.F. 1908. Note on the occurrence of *Cassiopea*

xamachana and *Polyclonia frondosa* at the Tortugas. *Papers from the Tortugas Laboratories* 1: 150–55.

Policansky, D. 1981. Sex choice and the size advantage model in jack-in-the-pulpit *(Arisaema triphyllum)*. *Proceedings of the National Academy of Sciences* 78: 1306–08.

Polis, G.A., and R.D. Farley. 1979. Behavior and ecology of mating in the cannibalistic scorpion, *Paruroctonus mesaensis* Stahnke (Scorpionida: Vaejovidae). *Journal of Arachnology* 7: 33–46.

Purcell, J.E. 1980. Influence of siphonophore behavior upon their natural diets: Evidence for aggressive mimicry. *Science* 209: 1045–47.

Raup, D.M., and J.J. Sepkoski, Jr. 1984. Periodicity of extinctions in the geologic past. *Proceedings of the National Academy of Sciences* 81: 801–05.

Reichler, J.L. 1981. *Fabulous baseball facts, feats and figures.* New York: Collier.

Reichler, J.L. 1982. *The baseball encyclopedia.* New York: MacMillan.

Roeder, K.D. 1935. An experimental analysis of the sexual behavior of the praying mantis *(Mantis religiosa* L.). *Biological Bulletin* 69: 203–20.

Ross, K., and R.L. Smith. 1979. Aspects of the courtship behavior of the black widow spider, *Laterodectus hesperus* (Araneae: Theridiidae), with evidence for the existence of a contact sex pheromone. *Journal of Arachnology* 7: 69–77.

Sagan, C. 1983. Nuclear war and climatic catastrophe: Some policy implications. *Foreign Affairs,* Winter 1983/84, pp. 257–92.

Schrödinger, E. 1944. *What is life?* Cambridge, England: Cambridge University Press.

Seilacher, A. 1984. Late Precambrian and early Cambrian Metazoa: Preservational or real extinctions. In Patterns of change in Earth evolution; ed. H.D. Holland and A.R. Trendall, pp. 159–68. Berlin: Springer Verlag.

Sepkoski, J.J.; R.K. Bambach; D.M. Raup; and J.W. Valentine. 1981. Phanerozoic marine diversity and the fossil record. *Nature* 293: 435.

Serres, E.R.A. 1833. Recherches d'anatomie transcendante

et pathéologique. Théorie des formations et des déformations organiques, appliquée à l'anatomie de Ritta Christina, et de la duplicité monstrueuse. *Mémoires de l'Académie Royale des Sciences* 11: 583–895.

Shapiro, D.Y. 1981. Sequence of coloration changes during sex reversal in the tropical marine fish *Anthias squamipinnis. Bulletin of Marine Sciences* 31: 383–98.

Siegel, R., quoted by J. Greenberg. 1983. Natural highs in natural habitats. *Science News,* November 5, pp. 300–01.

Simpson, G.G. 1964. The nonprevalence of humanoids. In *This view of life,* Essay 13, pp. 253–71. New York: Harcourt, Brace and World.

Sterba, G. 1973. *Freshwater fishes of the world.* Hong Kong TFH Publications, 2 vols.

Sulloway, F.J. 1982. Darwin and his finches: The evolution of a legend. *Journal of the History of Biology* 15: 1–53.

Sulloway, F.J. 1982. Darwin's conversion: The *Beagle* voyage and its aftermath. *Journal of the History of Biology* 15: 325–96.

Swainson, W. 1835. *On the natural history and classification of quadrupeds.* London: Longman, Rees, Orme, Brown, Green and Longman.

Thomson, W. (later Lord Kelvin). 1866. The "doctrine of uniformity" in geology briefly refuted. *Proceedings of the Royal Society of Edinburgh* 5: 512–13.

Tipler, F.J. 1981. Extraterrestrial intelligent beings do not exist. *Physics Today,* April, pp. 9, 70–71.

Tipler, F.J. 1982. We are alone in our galaxy. *New Scientist* 96 (October 7), pp. 33–35.

Tyson, E. 1699. *Orang-Outang,* sive Homo sylvestris; *or, the anatomy of a pygmie compared with that of a monkey, an ape, and a man.* London: Thomas Bennet.

Van Valen, L. 1974. A new evolutionary law. *Evolutionary Theory* 1: 1–30.

Wallace, A.R. 1870. The measurement of geological time. *Nature* 1: 399–401, 452–55.

Wallace, A.R. 1903. *Man's place in the universe: A study of the results of scientific research in relation to the unity or plurality of worlds.* New York: McClure Phillips and Company.

Wells, W.C. 1818. Account of a female of the white race of mankind, part of whose skin resembles that of a Negro. In *Two Essays: One upon single vision with two eyes, the other on dew.* London: Archibald Constable and Company.

White, C. 1799. *An account of the regular gradation in man, and in different animals and vegetables.* London: C. Dilly.

Whitmire, D.P., and A.A. Jackson IV. 1984. Are periodic mass extinctions driven by a distant solar companion? *Nature* 308: 713–15.

Index

FOR THE BEST IN PAPERBACKS, LOOK FOR THE

In every corner of the world, on every subject under the sun, Penguin represents quality and variety – the very best in publishing today.

For complete information about books available from Penguin – including Pelicans, Puffins, Peregrines and Penguin Classics – and how to order them, write to us at the appropriate address below. Please note that for copyright reasons the selection of books varies from country to country.

In the United Kingdom: For a complete list of books available from Penguin in the U.K., please write to *Dept E.P., Penguin Books Ltd, Harmondsworth, Middlesex, UB7 0DA*

In the United States: For a complete list of books available from Penguin in the U.S., please write to *Dept BA, Penguin, 299 Murray Hill Parkway, East Rutherford, New Jersey 07073*

In Canada: For a complete list of books available from Penguin in Canada, please write to *Penguin Books Canada Ltd, 2801 John Street, Markham, Ontario L3R 1B4*

In Australia: For a complete list of books available from Penguin in Australia, please write to the *Marketing Department, Penguin Books Australia Ltd, P.O. Box 257, Ringwood, Victoria 3134*

In New Zealand: For a complete list of books available from Penguin in New Zealand, please write to the *Marketing Department, Penguin Books (NZ) Ltd, Private Bag, Takapuna, Auckland 9*

In India: For a complete list of books available from Penguin, please write to *Penguin Overseas Ltd, 706 Eros Apartments, 56 Nehru Place, New Delhi, 110019*

In Holland: For a complete list of books available from Penguin in Holland, please write to *Penguin Books Nederland B.V., Postbus 195, NL–1380AD Weesp, Netherlands*

In Germany: For a complete list of books available from Penguin, please write to *Penguin Books Ltd, Friedrichstrasse 10 – 12, D–6000 Frankfurt Main 1, Federal Republic of Germany*

In Spain: For a complete list of books available from Penguin in Spain, please write to *Longman Penguin España, Calle San Nicolas 15, E–28013 Madrid, Spain*

FOR THE BEST IN PAPERBACKS, LOOK FOR THE 🐧

A CHOICE OF PENGUINS AND PELICANS

Metamagical Themas   Douglas R. Hofstadter

A new mind-bending bestseller by the author of *Gödel, Escher, Bach*.

The Body   Anthony Smith

A completely updated edition of the well-known book by the author of *The Mind*. The clear and comprehensive text deals with everything from sex to the skeleton, sleep to the senses.

Why Big Fierce Animals are Rare   Paul Colinvaux

'A vivid picture of how the natural world works' – *Nature*

How to Lie with Statistics   Darrell Huff

A classic introduction to the ways statistics can be used to prove *anything*, the book is both informative and 'wildly funny' – *Evening News*

The Penguin Dictionary of Computers   Anthony Chandor and others

An invaluable glossary of over 300 words, from 'aberration' to 'zoom' by way of 'crippled lead-frog tests' and 'output bus drivers'.

The Cosmic Code   Heinz R. Pagels

Tracing the historical development of quantum physics, the author describes the baffling and seemingly lawless world of leptons, hadrons, gluons and quarks and provides a lucid and exciting guide for the layman to the world of infinitesimal particles.

The Second World War (6 volumes)   Winston S. Churchill

The definitive history of the cataclysm which swept the world for the second time in thirty years.

1917: The Russian Revolutions and the Origins of Present-Day Communism
Leonard Schapiro

A superb narrative history of one of the greatest episodes in modern history by one of our greatest historians.

Imperial Spain 1496–1716   J. H. Elliot

A brilliant modern study of the sudden rise of a barren and isolated country to be the greatest power on earth, and of its equally sudden decline. 'Outstandingly good' – *Daily Telegraph*

Joan of Arc: The Image of Female Heroism   Marina Warner

'A profound book, about human history in general and the place of women in it' – Christopher Hill

Man and the Natural World: Changing Attitudes in England 1500–1800
Keith Thomas

'A delight to read and a pleasure to own' – Auberon Waugh in the *Sunday Telegraph*

The Making of the English Working Class   E. P. Thompson

Probably the most imaginative – and the most famous – post-war work of English social history.